SUPPORT OF UNDERGROUND EXCAVATIONS IN HARD ROCK

Support of Underground Excavations in Hard Rock

E. HOEK
Vancouver, B.C., Canada

P.K. KAISER
Geomechanics Research Centre, Laurentian University, Sudbury, Ont., Canada

W.F. BAWDEN
Department of Mining Engineering, Queen's University, Kingston, Ont., Canada

LONDON AND NEW YORK

Funding by Mining Research Directorate and Universities Research Incentive Fund

First print: 1995
Second print: 1997
Third print: 1998
Fourth print: 2005

Published by Taylor & Francis
2 Park Square, Milton Park, Abingdon, Oxon, OX14 4RN
270 Madison Ave, New York NY 10016

Transferred to Digital Printing 2006

ISBN 90-5410-186-5 (Hardback)
ISBN 90-5410-187-3 (Paperback)

Publisher's Note
The publisher has gone to great lengths to ensure the quality of this reprint but points out that some imperfections in the original may be apparent

Table of contents

Foreword

Support of underground excavations in hard rock is the most complete and up-to-date manual for use in the design of excavations and support mechanisms for underground mines.

This work resulted from close collaboration between industry and university in the fields of pre-competitive research. The mining industry provided funding and advisory support through the recently formed Mining Research Directorate (MRD). The Universities of Toronto, Laurentian and Queen's, under the direction of Professors Hoek, Kaiser and Bawden, provided the stimulus and facility for carrying out the tasks. These professors were assisted by some 40 engineers and graduate students who researched knowledge sources and experience world-wide.

The final product includes three computer programs: DIPS, UNWEDGE and PHASES. These programs were funded jointly by the mining industry, through the MRD, and by the Universities Research Incentive Fund.

This Canadian book and the associated programs should prove to be an invaluable contribution to the training of mining engineers and technologists at universities and colleges throughout the world, and should prove extremely useful to underground mining practitioners, everywhere. The overall focus is directed towards more productive, safer and environmentally sound mining operations.

The book *Support of underground excavations in hard rock* testifies to the willingness of Canadian industries and universities to collaborate in the field of pre-competitive research and learning, to jointly pursue excellence, and to work together towards the economic and social betterment of our society.

Walter Curlook

Dr. Walter Curlook
Inco Limited
May 18, 1993.

Preface

This volume is the product of four years of research carried out under the direction of Professor Evert Hoek of the Department of Civil Engineering at the University of Toronto, Professor Peter K. Kaiser of the Geomechanics Research Centre at Laurentian University and Professor William F. Bawden of the Department of Mining Engineering at Queen's University.

Funding was co-ordinated by the Mining Research Directorate and was provided by Belmoral Mines Limited, Cominco Limited, Denison Mines Limited, Falconbridge Limited, Hudson Bay Mining and Smelting Co. Ltd., Inco Limited, Lac Minerals Limited, Minnova Inc., Noranda Inc., Placer-Dome Inc., Rio Algom Limited and the Teck Corporation.

Funding was also provided by the University Research Incentive Fund, administered by the Ministry of Colleges and Universities of Ontario.

The results of this research are summarised in this book. The programs DIPS, UNWEDGE and PHASES, used in this book, were developed during the project and are available from the Rock Engineering Group of the University of Toronto[1].

Many individuals have contributed to the preparation of the book and the associated programs; it would be impossible to name them all. The advice and encouragement provided by the Technical Advisory Committee on the project and by Mr Charles Graham, Managing Director of MRD, are warmly acknowledged. The assistance provided by the many engineers and miners at the various field sites, on which research was carried out, was greatly appreciated. The major contributions made by research engineers and graduate students who were supported by this project are acknowledged. Special thanks are due to Dr Jean Hutchinson, who assisted with the writing of this volume and the program manuals, to Dr José Carvalho, Mr Mark Diederichs, Mr Brent Corkum and Dr Bin Li who were responsible for most of the program development and to Mrs Theo Hoek who compiled the list of references and proof-read the final manuscript.

A draft of the book was sent to a number of reviewers around the world. Amost all of them responded, contributing very constructive criticisms and suggestions. As a result of this review process, several chapters were re-written and one new chapter was added. While it would not be practical to list all of these reviewers individually, the authors wish to express their sincere thanks to all those who took so much trouble to review the draft and whose contributions have added to the value of the book.

The authors anticipate that some of the subject matter contained in this book will be superseded quite quickly as the technology of

---

[1] An order form for these programs is included at the back of this book.

underground support continues to develop. Readers are encouraged to send comments or suggestions which can be incorporated into future editions of the book. These contributions can be sent to any one of the authors at the addresses listed below.

Dr Evert Hoek
West Broadway Professional Centre
412-2150 West Broadway
Vancouver, British Columbia, V6K 4L9
Canada

Professor Peter Kaiser
Geomechanics Research Centre
Laurentian University
Fraser Building F217
Ramsey Lake Road
Sudbury, Ontario P3E 2C6
Canada

Professor W.F. Bawden
Department of Mining Engineering
Goodwin Hall
Queen's University
Kingston, Ontario K7L 3N6
Canada

1 An overview of rock support design

1.1 Introduction

The potential for instability in the rock surrounding underground mine openings is an ever-present threat to both the safety of men and equipment in the mine. In addition, because of dilution of the ore due to rock falls, the profitability of the mining operation may be reduced if failures are allowed to develop in the rock surrounding a stope. In order to counteract these threats, it is necessary to understand the causes of the instability and to design measures which will eliminate or minimise any problems.

It is important to recognise that there are two scales involved in the creation of potential instability problems. The first scale, which may be termed the mine scale, is one involving the entire orebody, the mine infrastructure and the surrounding rock mass. The second or local scale is limited to the rock in the vicinity of the underground openings. These two scales are illustrated in Figure 1.1.

The composition and nature of the orebody and the surrounding host rock, the in situ stresses and the geometry and excavation sequence of the stopes, all have an influence upon the overall stability of the mine. Mining stopes in the incorrect sequence, leaving pillars of inadequate size between stopes, incorrectly locating shafts and orepasses, in areas which are likely to be subjected to major stress changes, are all problems which have to be dealt with by considering the overall geometry of the mine.

On the other hand, the stability of the rock surrounding a single stope, a shaft station or a haulage depends on stress and structural conditions in the rock mass within a few tens of metres of the opening boundary. The local stresses are influenced by the mine scale conditions, but local instability will be controlled by local changes in stress, by the presence structural features and by the amount of damage to the rock mass caused by blasting. In general, it is the local scale which is of primary concern in the design of support.

1.2 Stages in mine development

Table 1.1 gives a summary of the different stages of mine development. Different amounts of information are available at each stage and this influences the approach to support design which can be used for each stage. Each of these stages is reviewed briefly in the following discussion. The reader should not be concerned if some of the terms or concepts included in this brief review are unfamiliar. These will be discussed in detail in later chapters of this volume.

It is also worth pointing out that the term 'support' is used to cover all types of rockbolts, dowels, cables, mesh, straps, shotcrete and steel sets used to minimise instability in the rock around the mine openings. In more detailed discussions in later chapters, terms

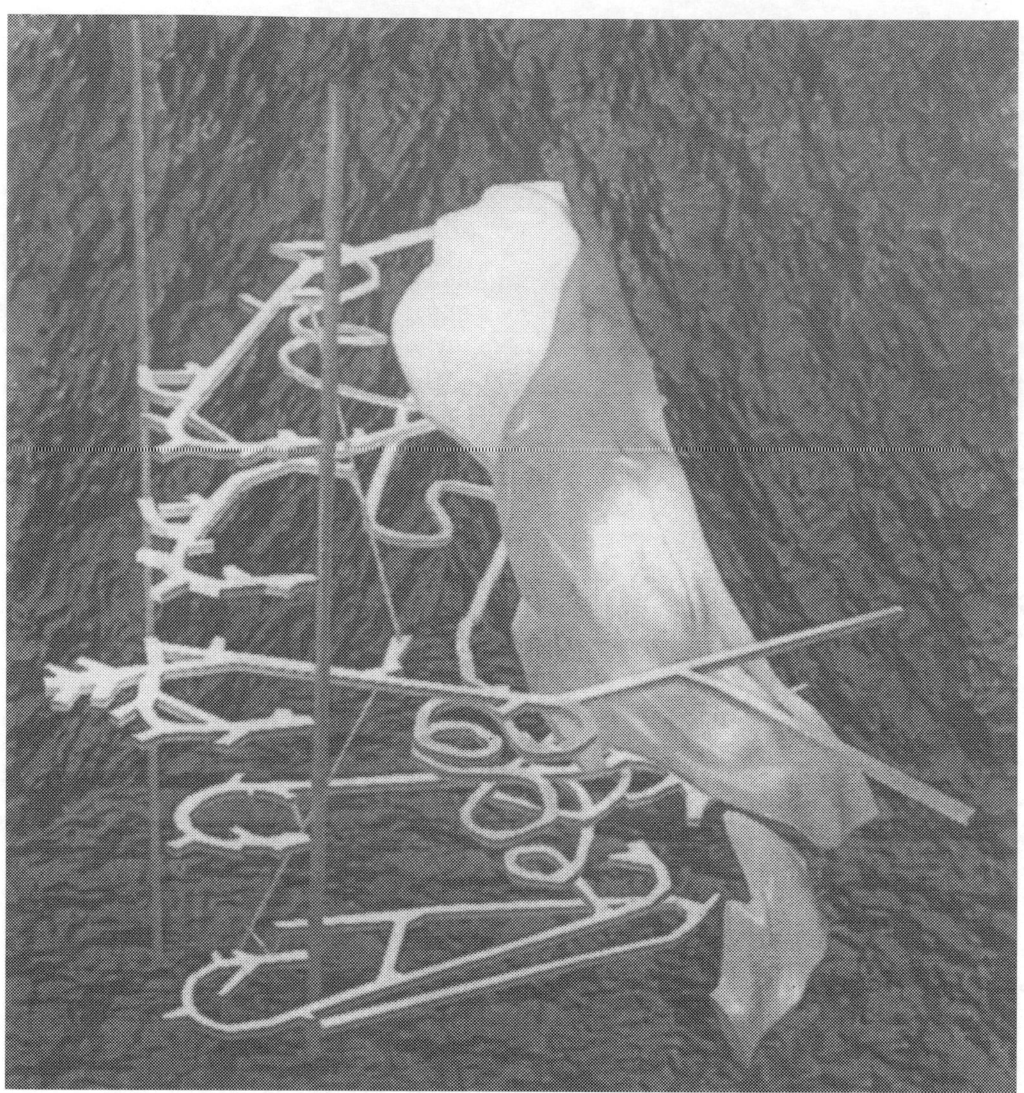

Figure 1.1: Underground instability problems are controlled by the overall geometry of the mine (upper image) and by in situ stresses and rock mass characteristics around each opening (lower photograph).
(Graphical image of mine created by Dr Murray Grabinsky of the Department of Civil Engineering, University of Toronto).

such as active support, passive support and reinforcement will be introduced to differentiate between the ways in which each of these support types works.

1.2.1 *Exploration and preliminary design*

The amount of information, which is available during the exploration and preliminary design stages of a mine, is usually limited to that obtained from regional geology maps, geophysical studies, surface mapping and exploration boreholes. Exploration drilling programmes generally do not include provision for obtaining geotechnical information and hence the information available from the boreholes may be limited to rock types and ore grades. Consequently, it is only possible to construct a very crude rock mass classification upon which preliminary estimates of rock support requirements can be based. This is generally not a major problem at this stage since the mine owner only needs to make a rough estimate of potential support costs.

More detailed estimates normally require the drilling of a few judiciously positioned boreholes and having a geotechnical technician carefully log the core. The information obtained from such an investigation is used to construct a rock mass classification and, possibly, to provide input for very simple numerical analyses. It can also provide a sound basis for planning more detailed site investigations for the next stage of the mine development.

1.2.2 *Mine design*

Once it has been concluded that the ore deposit can be mined profitably and an overall mining strategy has been developed, the project can move into the next stage which usually involves sinking an exploration shaft or, if the orebody is shallow, a ramp and exploration adits. These provide underground access to the ore body and the surrounding rock mass and also permit much more detailed geotechnical evaluation than was possible during the exploration stage.

Structural mapping of the features exposed in the exploration openings, laboratory testing of core samples obtained from underground drilling and measurement of in situ stresses are the types of activities which should be included in the geotechnical programme associated with this stage. Observations of the rock mass failure can be used to estimate rock mass properties and in situ stresses. These activities also provide information for the construction of rock mass classifications and for numerical models, which can be used for the preliminary analysis of instability around typical mine openings.

Studies carried out during the mine design stage can also be used to estimate the support requirements for permanent openings such as shaft stations, refuge stations, underground crusher chambers, ramps and haulages. These designs tend to be more conservative than those for the support in normal mine openings, since safety of men and equipment is a prime consideration in these permanent openings.

An important activity at this stage of the mine development programme is the layout of stopes and the choice of stope dimensions and stoping sequence. The role of support and of backfilling the

Table 1.1: Summary of information available and approaches to the design of support at various stages of mine development.

Development stage	Information available	Design approach
Exploration and preliminary design	Location and nature of ore deposit and rudimentary information on host rock from exploration drilling and surface mapping	Use rock mass classification to obtain first estimates of support requirements
Mine design with detailed layout of permanent access excavations and stopes	Estimates of rock mass structure and strength and in situ stress conditions from mapping and measurements carried out in exploration adits and shafts	Use modified rock mass classifications and numerical models to design rockbolt patterns for permanent excavations and cable bolt patterns for stopes
Mining shafts, haulages, ramps and other permanent access ways and early stopes of mine	Detailed information on rock mass structure and strength, blast damage, and on the performance of selected support systems	Refinement of designs using elaborate numerical models and giving attention to quality control of support installation
Later years of mining and extraction of pillars	Experience from many years of refinement of excavation and support installation and, possibly, the results of displacement and load monitoring	Fine tuning of support to meet specific requirements with the use of the most advanced numerical models where required

stopes have to be evaluated. In entry stopes, such as cut and fill stopes, the support is required for both safety and dilution control. The primary function of support in non-entry stopes is for the control of dilution.

1.2.3 *Early years of mining*

During the early years of mining, a significant amount of effort will be devoted to excavating and stabilising the permanent mine openings such as shafts, shaft stations, haulages, ramps, orepasses, underground crusher chambers, underground garages, electrical substations and refuge stations. These excavations are required to provide safe access for the life of the mine or for a significant part of its life and, hence, a high degree of security is required. The design of these excavations is similar, in many ways, to the design of civil engineering tunnels and caverns and a high density of support may be required in order to reduce potential instability to an absolute minimum. What separates the support of mining openings from the support of similar civil engineering structures is the fact that mine openings may have to survive large deformations as a result of changing stress conditions induced by progressive mining. The support has to remain effective in gradually degrading rock, and it may have to sustain dynamic loads.

The design of this support requires a fairly detailed knowledge of the rock mass structure and the in situ stress conditions. These are generally obtained from the geotechnical studies associated with the mine design stage discussed earlier. Numerical models can be used to estimate the extent of potential instability in the rock surrounding permanent mine openings and to design typical support systems to control this instability. In general, the design of support for these permanent openings tends to be conservative in that the designer will generally err on the side of specifying more, rather than less support, to take care of unforeseen conditions. Rehabilitation of permanent openings can disrupt mine operations and can be difficult and expensive. Consequently, the aim is to do the job once and not have to worry about it again. Special methods of monitoring the rock mass response may be justified for back analysis studies aimed at improving the understanding of the support performance.

At this stage of the development of the mine, the stopes will generally be relatively small and isolated and it should be possible to maintain safety and minimise dilution with a modest amount of support. However, it is important that stress changes, which will be associated with later mining stages, be anticipated and provision made for dealing with them. This may mean that support has to be placed in areas which appear to be perfectly stable in order to preserve the stability of the rock during later mining.

A good example of this type of pre-placed support can be found in the reinforcement of drawpoints. When these are mined, before the stope above them has been blasted, they are generally in stable rock which does not require support. However, when the overlying stopes are blasted and the drawpoints come into operation, the stress changes, due to the creation of a large new excavation and the dynamic forces resulting from the movement of the broken ore, can result in serious overstressing of the rock surrounding the drawpoint. Where these changes have been anticipated and this rock mass has been reinforced, the stability of the drawpoints can be maintained for the life of the stope to which they provide access. Typically, untensioned grouted wire rope, installed during excavation of the trough drive and the drawpoints, provides excellent reinforcement for these conditions. Wire rope is recommended in place of steel rebar because of its greater resistance to damage due to impact from large rocks.

This stage of mining also provides an opportunity to sort out some of the practical problems associated with support installation. For example, the water-cement ratio of the grout used for grouting cables in place is an important factor in determining the capacity of this type of support. Pumping a low water-cement ratio grout requires both the correct equipment and a well-trained crew. Investment of the time and effort required to sort out equipment problems and to train the crews will be amply rewarded at later stages in the mine development.

1.2.4 *Later years of mining*

When an underground mine reaches maturity and the activities move towards the mining of stopes of significant size and the recovery of pillars, the problems of support design tend to become severe. The mine engineer is now required to use all of the experience, gained in the early trouble-free years of mining, to design support systems which will continue to provide safe access and to minimise dilution.

Depending upon the nature and the scale of the potential instability problems encountered, the support may be similar to that used earlier, or new and innovative support designs may be implemented. It is generally at this time that the use of the most sophisticated support design techniques is justified.

At this stage of mining a good geotechnical database should be available. This may include the results of observations and measurements of excavation deformation, rock mass failure, support performance and in situ stress changes. An analysis of these measurements and observations can provide a sound basis for estimating the

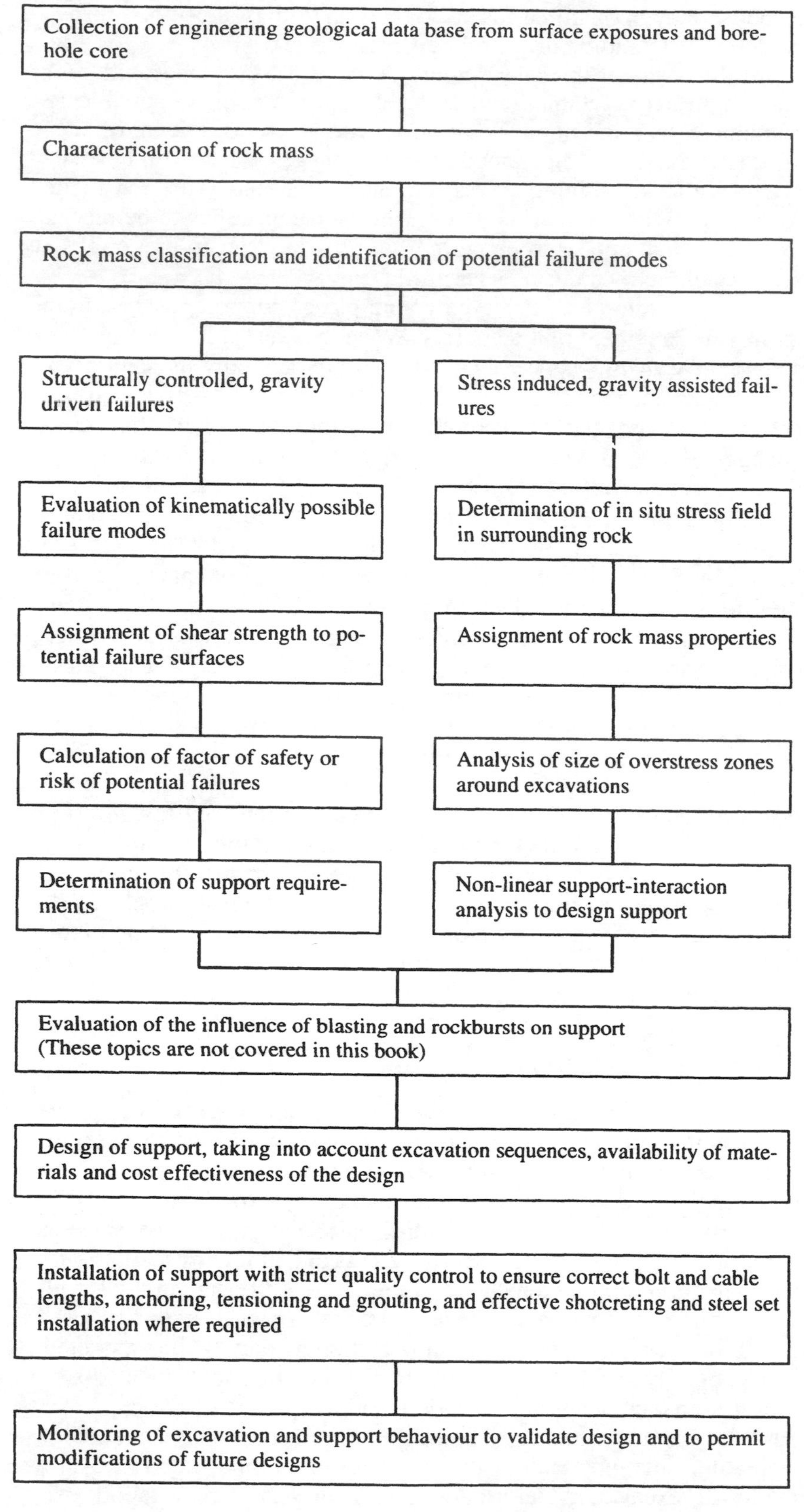

Figure 1.2: Steps involved in support design for underground excavations in hard rock.

future behaviour of stopes and pillars and for designing support systems to stabilise the openings.

1.3 Support design

While the amount of information available at various stages of mine design, development and production varies, the basic steps involved in the design of support remain unchanged. The lack of information at the early stages of mine design and development means that some of the steps in this design process may have to be skipped or be based upon rough estimates of the structural geology, in situ stresses, rock mass strength and other information.

The basic steps involved in the design of support for underground hard rock mines are summarised in Figure 1.2.

2 Assessing acceptable risks in design

2.1 Introduction

How does one assess the acceptability of an engineering design? Relying on judgement alone can lead to one of the two extremes illustrated in Figure 2.1. The first case is economically unacceptable while the example illustrated in the lower drawing violates all normal safety standards.

Figure 2.1: Rockbolting alternatives based on individual judgement. (Drawings from a cartoon in a brochure on rockfalls published by the Department of Mines of Western Australia.)

2.2 Factor of safety

The classical approach used in designing engineering structures is to consider the relationship between the capacity C (strength or resisting force) of the element and the demand D (stress or disturbing force). The Factor of Safety of the structure is defined as $F = C/D$ and failure is assumed to occur when F is less than 1.

Consider the case of a pattern of rockbolts which are designed to hold up a slab of rock in the back of an excavation. Figure 2.2 shows a slab of thickness t being supported by one rockbolt in a pattern spaced on a grid spacing of $S \times S$. Assuming that the unit weight of the broken rock is $\gamma = 2.7$ tonnes/m^3, the thickness of the slab $t = 1$ m and that grid spacing $S = 1.5$ m, the weight of the block being carried by the bolt, is given by $W = \gamma.t.S^2 = 6.1$ tonnes. The demand D on the rockbolt is equal to the weight W of the block and, if the strength or capacity of the bolt is $C = 8$ tonnes, the factor of safety $F = 8/6.1 = 1.3$.

The value of the factor of safety, which is considered acceptable for a design, is usually established from previous experience of successful designs. A factor of safety of 1.3 would generally be considered adequate for a temporary mine opening while a value of 1.5 to 2.0 may be required for a 'permanent' excavation such as an underground crusher station.

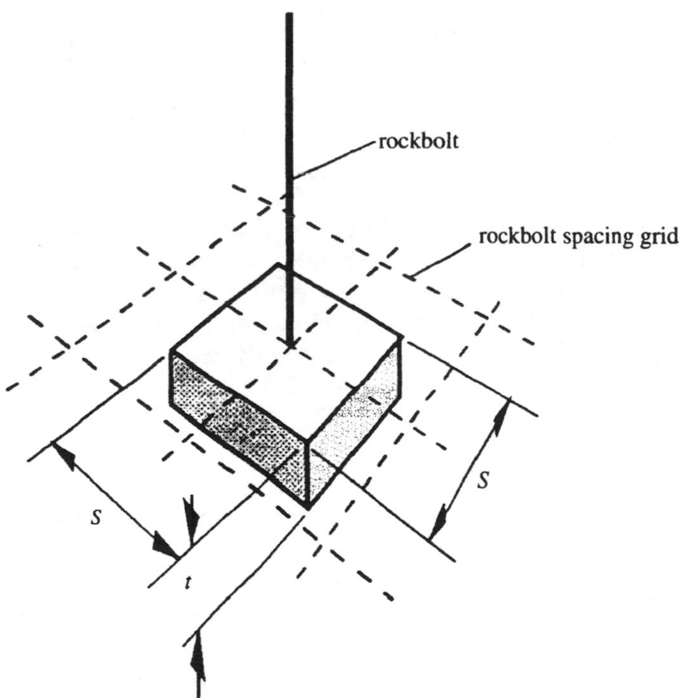

Figure 2.2: Roof slab of thickness t being supported by a rockbolt in a pattern spaced on a grid of $S \times S$.

2.3 Sensitivity studies

Rather than base an engineering design decision on a single calculated factor of safety an alternative approach, which is frequently used to give a more rational assessment of the risks associates with a particular design, is to carry out a sensitivity study. This involves a series of calculations, in which each significant parameter is varied systematically over its maximum credible range, in order to determine its influence upon the factor of safety.

In the very simple example discussed in the previous section, the rockbolt designer may consider that the thickness t of the slab could vary from 0.7 to 1.3 m and that the strength of the rockbolts could lie between 7 and 9 tonnes. Hence, keeping all other parameters constant, the factor of safety will vary from $7/(2.7 \times 1.3 \times 1.5^2) = 0.88$ to a maximum of $9/(2.7 \times 0.7 \times 1.5^2) = 2.12$.

The minimum factor of safety of 0.88 is certainly unacceptable and the designer would then have to decide what to do next. If it was felt that a significant number of bolts could be overloaded, common sense would normally dictate that the average factor of safety of 1.3 should be increased to say 1.5 by decreasing the bolt spacing from 1.5 to 1.4 m. This would give a minimum factor of safety of 1.02 and a maximum of 2.43 for the assumed conditions.

2.4 The application of probability to design

The very simple sensitivity study discussed above is the type of calculation which is carried out routinely on sites around the world. In an on-going mining operation the number of rockbolt failures would soon indicate whether the average design was acceptable or whether modifications were required.

It will be evident to the reader that this design process involves a considerable amount of judgement based upon experience built up from careful observations of actual performance. When no such experience is available because the design is for a new area or for a new mine, what tools are available to assist the designer in making engineering decisions? While the use of probability theory does not provide all the answers which the designer may seek, it does offer a means for assessing risk in a rational manner, even when the amount of data available is very limited.

A complete discussion on probability theory exceeds the scope of this book and the techniques discussed on the following pages are intended to introduce the reader to the subject and to give an indication of the power of these techniques in engineering decision making. A more detailed treatment of this subject will be found in a book by Harr (1987) entitled 'Reliability-based design in civil engineering'. A paper on geotechnical applications of probability theory entitled 'Evaluating calculated risk in geotechnical engineering' was published by Whitman (1984) and is recommended reading for anyone with a serious interest in this subject. Pine (1992), Tyler et al. (1991), Hatzor and Goodman (1992) and Carter (1992) have published papers on the application of probability theory to the analysis of problems encountered in underground mining and civil engineering.

Most geotechnical engineers regard the subject of probability theory with doubt and suspicion. At least part of the reason for this mistrust is associated with the language which has been adopted by those who specialise in the field of probability theory and risk assessment. The following definitions are given in an attempt to dispel some of the mystery which tends to surround this subject.

Random variables: Parameters such as the angle of friction of rock joints, the uniaxial compressive strength of rock specimens, the inclination and orientation of discontinuities in a rock mass and the measured in situ stresses in the rock surrounding an opening do not have a single fixed value, but may assume any number of values. There is no way of predicting exactly what the value of one of these parameters will be at any given location. Hence these parameters are described as random variables.

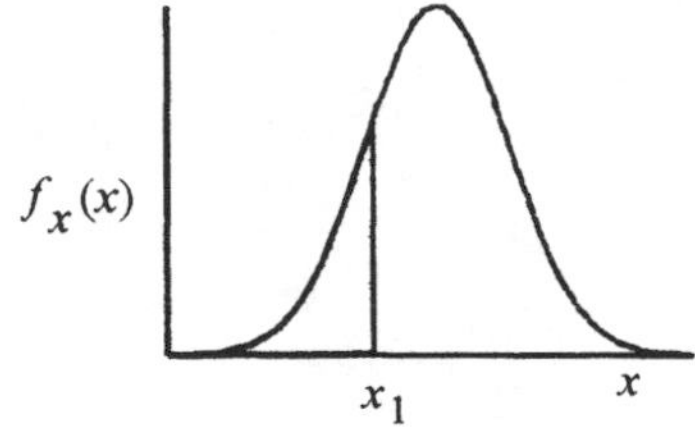

Probability density function (PDF)

Probability distribution: A probability density function (PDF) describes the relative likelihood that a random variable will assume a particular value. A typical probability density function is illustrated opposite. In this case the random variable is continuously distributed (i.e., it can take on all possible values). The area under the PDF is always unity.

An alternative way of presenting the same information is in the form of a cumulative distribution function (CDF) which gives the probability that the variable will have a value less than or equal to the selected value. The CDF is the integral of the corresponding probability density function, i.e., the ordinate at x_1, on the cumulative distribution, is the area under the probability density function to the left of x_1. Note the $f_x(x)$ is used for the ordinate of a PDF while $F_x(x)$ is used for a CDF.

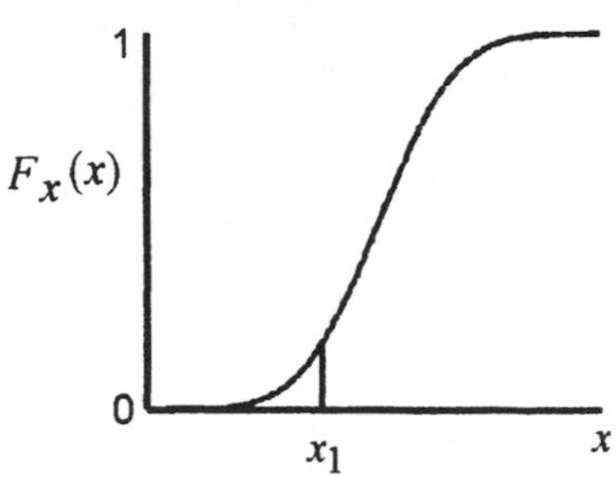

Cumulative distribution function (CDF)

One of the most common graphical representations of a probability distribution is a histogram in which the fraction of all observations, falling within a specified interval, is plotted as a bar above that interval.

Data analysis: For many applications it is not necessary to use all of the information contained in a distribution function. Quantities, summarised only by the dominant features of the distribution, may be adequate.

The *sample mean* or *expected value* or *first moment* indicates the centre of gravity of a probability distribution. A typical application would be the analysis of a set of results $x_1, x_2, \ldots\ldots\ldots, x_n$ from uniaxial strength tests carried out in the laboratory. Assuming that there are n individual test values x_i, the mean $\bar{x}$ is given by:

$$\bar{x} = \frac{1}{n}\sum_{i=1}^{n} x_i \tag{2.1}$$

The *sample variance* s^2 or the *second moment about the mean* of a distribution is defined as the mean of the square of the difference between the value of x_i and the mean value $\bar{x}$. Hence:

$$s^2 = \frac{1}{n-1}\sum_{i=1}^{n}(x_i - \bar{x})^2 \tag{2.2}$$

Note that, theoretically, the denominator for calculation of variance of samples should be n, not (n-1). However, for a finite number of samples, it can be shown that the correction factor $n/(n-1)$, known as Bessel's correction, gives a better estimate. For practical purposes the correction is only necessary when the sample size is less than 30.

The *standard deviation* s is given by the positive square root of the variance s^2. In the case of the commonly used normal distribution, about 68% of the test values will fall within an interval defined by the *mean* ± *one standard deviation* while approximately 95% of all the test results will fall within the range defined by the *mean* ± *two standard deviations.* A small standard deviation will indicate a tightly clustered data set, while a large standard deviation will be found for a data set in which there is a large scatter about the mean.

The *coefficient of variation* (COV) is the ratio of the standard deviation to the mean, i.e. COV = $s/\bar{x}$. COV is dimensionless and it is a particularly useful measure of uncertainty. A small uncertainty would typically be represented by a COV = 0.05 while considerable uncertainty would be indicated by a COV = 0.25.

Normal distribution: The *normal* or *Gaussian* distribution is the most common type of probability distribution function and the distributions of many random variables conform to this distribution. It is generally used for probabilistic studies in geotechnical engineering unless there are good reasons for selecting a different distribution. Typically, variables which arise as a sum of a number of random effects, none of which dominate the total, are normally distributed.

The problem of defining a normal distribution is to estimate the values of the governing parameters which are the true mean (μ) and true standard deviation (σ). Generally, the best estimates for these values are given by the sample mean and standard deviation, determined from a number of tests or observations. Hence, from equations 2.1 and 2.2:

$$\mu = \bar{x} \tag{2.3}$$

$$\sigma = s \tag{2.4}$$

It is important to recognise that equations 2.3 and 2.4 give the most probable values of μ and σ and not necessarily the true values.

Obviously, it is desirable to include as many samples as possible in any set of observations. However, in geotechnical engineering, there are serious practical and financial limitations to the amount of data which can be collected. Consequently, it is often necessary to make estimates on the basis of judgement, experience or from comparisons with results published by others. These difficulties are often used as an excuse for not using probabilistic tools but, as will be shown later in this chapter, useful results can still be obtained from very limited data.

Having estimated the mean μ and standard deviation σ, the probability density function for a normal distribution is defined by:

$$f_x(x) = \frac{\exp\left[-\frac{1}{2}\left(\frac{x-\mu}{\sigma}\right)^2\right]}{\sigma\sqrt{2\pi}} \tag{2.5}$$

for $-\infty \leq x \leq \infty$.

As will be seen later, this range from $-\infty$ to $+\infty$ can cause problems when a normal distribution is used as a basis for a Monte Carlo analysis in which the entire range of values is randomly sampled. This can give rise to a few very small (sometimes negative) and very large numbers which, in some cases, can cause numerical instability. In order to overcome this problem, the normal distribution is sometimes truncated so that only values falling within a specified range are considered valid.

The cumulative distribution function (CDF) of a normal distribution must be found by numerical integration since there is no closed form solution.

Other distributions: In addition to the commonly used normal distribution, there are a number of alternative distributions which are used in probability analyses. Some of the most useful are:

- *Beta* distributions (Harr, 1987) are very versatile distributions, which can be used to replace almost any of the common distributions and which do not suffer from the extreme value problems discussed above, because the domain (range) is bounded by specified values.
- *Exponential* distributions are sometimes used to define events such as the occurrence of earthquakes or rockbursts or quantities such as the length of joints in a rock mass.
- *Lognormal* distributions are useful when considering processes such as the crushing of aggregates in which the final particle size results from a number of collisions of particles of many sizes, moving in different directions with different velocities. Such multiplicative mechanisms tend to result in variables which are lognormally distributed as opposed to the normally distributed variables resulting from additive mechanisms.
- *Weibul* distributions are used to represent the lifetime of devices in reliability studies or the outcome of tests, such as point load tests on rock core, in which a few very high values may occur.

It is no longer necessary for the person starting out in the field of probability theory to know and understand the mathematics involved in all of these probability distributions. Today, commercially available software programs can be used to carry out many of the computations automatically. Note that the authors are not advocating the blind use of 'black-box' software and the reader should exercise extreme caution in using such software without trying to understand exactly what the software is doing. However, there is no point in writing reports by hand if one is prepared to spend the time learning how to use a good word-processor correctly and the same applies to mathematical software.

One of the most useful software packages for probability analysis is a program called BestFit[1]. It has a built-in library of 18 probability distributions and it can be used to fit any one of these distributions to

---

[1] BestFit for Windows and its companion program @RISK for Microsoft Excel or Lotus 1-2-3 (for Windows or Macintosh) are available from the Palisade Corporation, 31 Decker Road, Newfield, New York 14867, USA. Fax number 1 607 277 8001.

a given set of data. Alternatively, it can be allowed automatically to determine the ranking of the fit of all 18 distributions to the data set. The results from such an analysis can be entered directly into a companion program called @RISK which can be used for risk evaluations using the techniques described below.

Sampling techniques: Consider the case of the rockbolt holding up a roof slab, illustrated in Figure 2.2. Assuming that the rockbolt spacing S is fixed, the slab thickness t and the rockbolt capacity C can be regarded as random variables. Assuming that the values of these variables are distributed about their means in a manner which can be described by one of the continuous distribution functions, such as the normal distribution described earlier, the problem is how to use this information to determine the distribution of factor of safety values.

The *Monte Carlo* method uses random or pseudo-random numbers to sample from probability distributions and, if sufficiently large numbers of samples are generated and used in a calculation, such as that for a factor of safety, a distribution of values for the end-product will be generated. The term 'Monte Carlo' is believed to have been introduced as a code word to describe this hit-and-miss sampling technique used during work on the development of the atomic bomb during World War II (Harr, 1987). Today Monte Carlo techniques can be applied to a wide variety of problems involving random behaviour and a number of algorithms are available for generating random Monte Carlo samples from different types of input probability distributions. With highly optimised software programs such as @RISK, problems involving relatively large samples can be run efficiently on most desktop or portable computers.

The *Latin Hypercube* sampling technique is a relatively recent development, which gives comparable results to the Monte Carlo technique, but with fewer samples (Imam et al., 1980, Startzman and Watterbarger, 1985). The method is based upon stratified sampling with random selection within each stratum. Typically, an analysis using 1000 samples obtained by the Latin Hypercube technique will produce comparable results to an analysis using 5000 samples obtained using the Monte Carlo method. This technique is incorporated into the program @RISK.

Note that both the Monte Carlo and the Latin Hypercube techniques require that the distribution of all the input variables should either be known or that they be assumed. When no information on the distribution is available, it is usual to assume a normal or a truncated normal distribution.

The *Generalised Point Estimate Method*, originally developed by Rosenbleuth (1981) and discussed in detail by Harr (1987), can be used for rapid calculation of the mean and standard deviation of a quantity, such as a factor of safety, which depends upon random behaviour of input variables. Hoek (1989) discussed the application of this technique to the analysis of surface crown pillar stability while Pine (1992) and Nguyen and Chowdhury (1985) have applied this technique to the analysis of slope stability and other mining problems.

To calculate a quantity, such as a factor of safety, two point estimates are made at one standard deviation on either side of the mean

($\mu \pm \sigma$) from each distribution representing a random variable. The factor of safety is calculated for every possible combination of point estimates, producing 2^n solutions, where n is the number of random variables involved. The mean and the standard deviation of the factor of safety are then calculated from these 2^n solutions.

While this technique does not provide a full distribution of the output variable, as do the Monte Carlo and Latin Hypercube methods, it is very simple to use for problems with relatively few random variables and is useful when general trends are being investigated. When the probability distribution function for the output variable is known, for example, from previous Monte Carlo analyses, the mean and standard deviation values can be used to calculate the complete output distribution. This was done by Hoek (1989) in his analysis of surface crown pillar failure.

2.5 Probability of failure

Considering again the very simple example of the roof slab supported by a pattern of rockbolts, illustrated in Figure 2.2, the following discussion illustrates the application of the probability techniques outlined above to the assessment of the risk of failure.

Table 2.1 lists the hypothetical results obtained from 62 pull out tests on 17 mm diameter expansion shell rockbolts with a nominal pull out strength of 8 tonnes. Figure 2.3a gives these results in the form of a frequency distribution or histogram. Each cross-hatched bar has been drawn so that its area is proportional to the number of values in the interval it represents. The continuous line on this plot represents a normal distribution which has been fitted to the input data using the program BestFit. This fitting process yields a mean or expected value for the pull-out tests as $\bar{C} = 7.85$ tonnes with a standard deviation of $\sigma = 0.37$. Note that the minimum and maximum values are 6.95 and 8.62 tonnes respectively. The cumulative probability distribution function for the same data set is given in Figure 2.3b.

The average thickness $\bar{t}$ of the roof slab being supported has been estimated at 1 m. Short of drilling dozens of holes to measure the variation in the value of t over a representative area of the roof, there is no way of determining a distribution for this variable in the same way as was possible for the rockbolt capacity. This is a common problem in geotechnical engineering, where it may be extremely difficult or even impossible to obtain reliable information on certain variables, and the only effective solution is to use educated guesswork.

In the case of the roof slab, it would not be unreasonable to assume that the thickness t is normally distributed about the mean of $\bar{t} = 1$ m. Obviously, t cannot be less than 0 since negative values produce a meaningless negative factor of safety while $t = 0$ results in 'divide by zero' errors. In order to avoid this problem the normal distribution has to be truncated. An arbitrary minimum value of $t = 0.25$ m has been used to truncate the lower end of the normal distribution since smaller values will produce very high factors of safety. It is unlikely that t would exceed say 2 m, and hence, this can

Table 2.1: Results of 62 pull-out tests on 17 mm diameter mechanically anchored rockbolts. (Units are tonnes).

6.95	7.01	7.15	7.23	7.31	7.41	7.42	7.44
7.48	7.48	7.54	7.54	7.55	7.61	7.63	7.64
7.64	7.65	7.66	7.67	7.69	7.71	7.73	7.73
7.75	7.75	7.75	7.78	7.78	7.8	7.8	7.81
7.85	7.86	7.86	7.88	7.91	7.93	7.93	7.94
7.97	7.99	8.02	8.02	8.03	8.03	8.05	8.1
8.12	8.13	8.19	8.21	8.23	8.23	8.23	8.25
8.26	8.3	8.31	8.34	8.48	8.62		

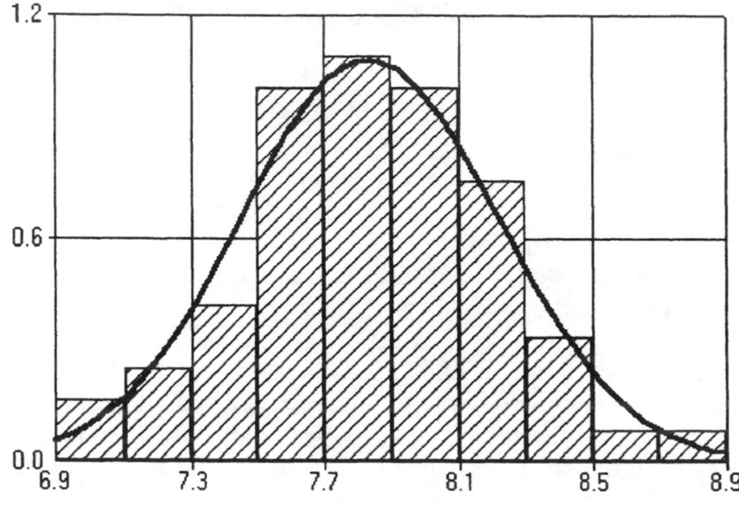

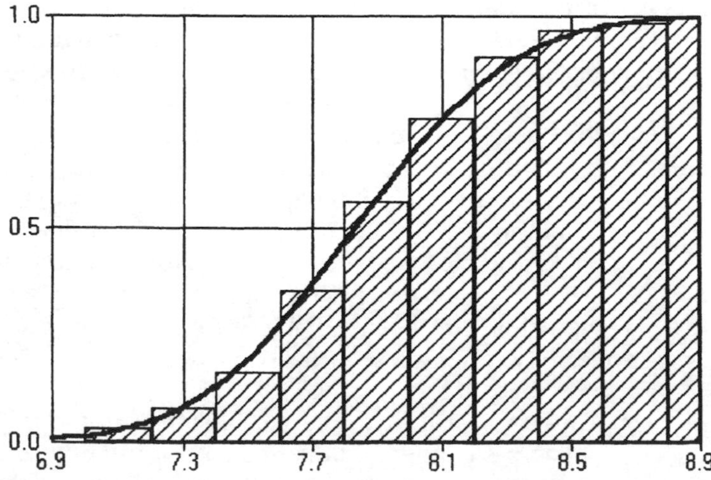

Figure 2.3: Hypothetical results from 62 pull-out tests on 17 mm diameter mechanically anchored rockbolts. The test results are plotted as histograms while fitted normal probability distributions are shown as continuous lines for a) a probability density function and b) a cumulative distribution function.

be used as an upper limit for the truncated normal distribution. For want of any better information it will be assumed that the standard deviation for the slab thickness is $\sigma = 0.5$ m. In other words, 68% of the slabs will be between 0.5 and 1.5 m thick while the remainder will be either thicker or thinner. Using these values to calculate the demand D produces a truncated normal distribution with minimum and maximum values of 1.52 tonnes and 12.15 tonnes respectively, a mean of 6.15 tonnes and a standard deviation of 2.82 tonnes.

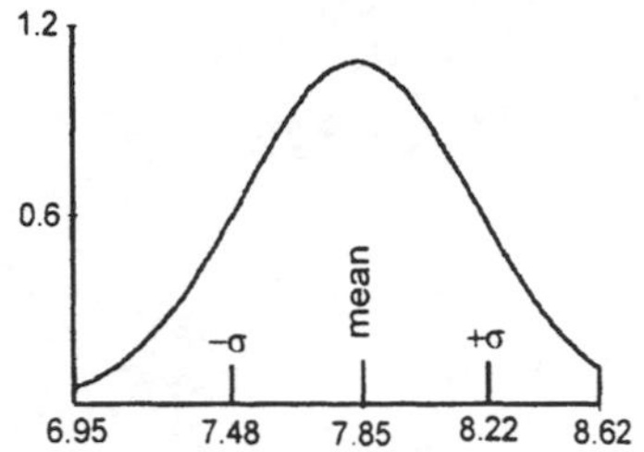

Rockbolt capacity C

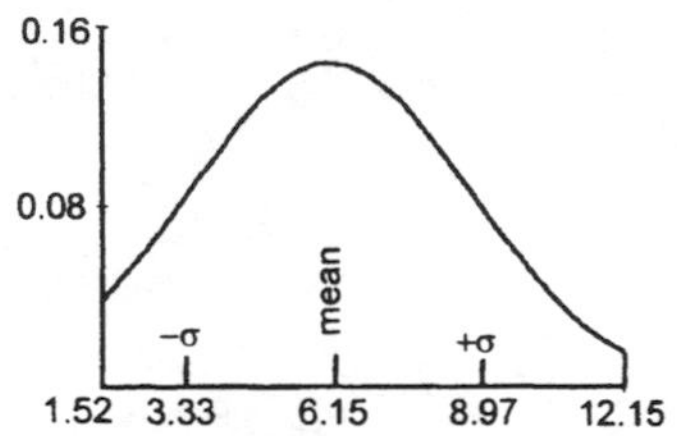

Demand D

Using the @RISK add-in program in a Microsoft Excel spreadsheet, the two truncated normal distributions illustrated in the margin drawings, representing the bolt capacity C and the load demand D, were each sampled 1000 times by means of the Latin Hypercube technique. The resulting pairs of samples were each used to calculate a factor of safety $F = C/D$. The resulting distribution of factors of safety is illustrated in Figure 2.4 which shows that a Lognormal distribution, defined by a mean of 1.41 and a standard deviation of 0.71, gives an adequate representation of the distribution. From the statistical records produced by @RISK, it was determined that approximately 30% of the 1000 cases sampled have factors of safety of less than 1.00, i.e., the *probability of failure* of this rockbolt design is 30% for the assumed conditions.

In order to establish whether a 30% probability of failure is acceptable, consider the consequences of one bolt in a pattern failing. The closest four bolts to this failed bolt would suddenly be called upon to carry an additional load of 20 to 25% over the load which they are already carrying. This is equivalent to increasing the bolt

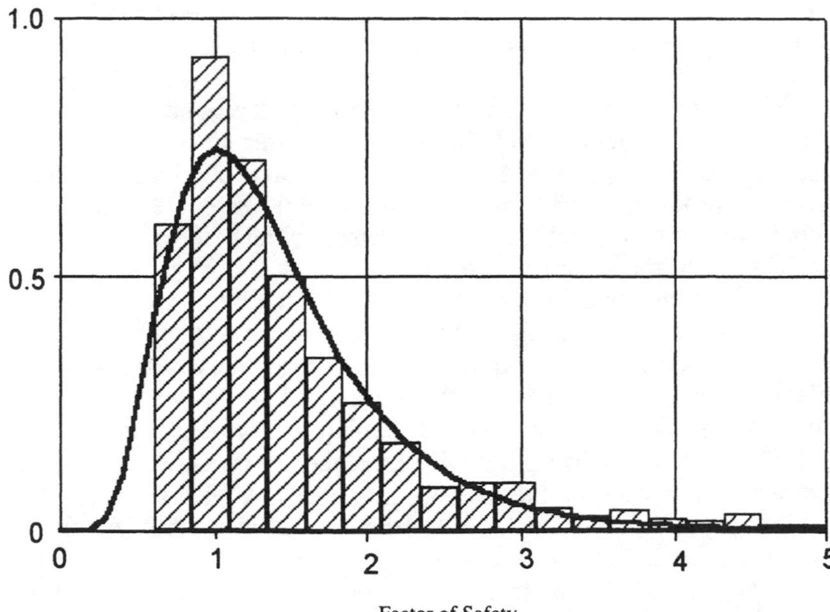

Figure 2.4: Lognormal distribution of factors of safety for a pattern of rockbolts supporting a roof slab. The distribution of factors of safety calculated by means of the Latin Hypercube technique are shown as a histogram while the fitted Lognormal distribution is shown as a continuous line.

spacing to about 1.65 m, and substitution of this value back into the @RISK analysis shows that the probability of failure increases to about 50%. This suggests that an expanding domino type failure process could occur and that the original factor of safety is not adequate.

Decreasing the bolt grid spacing to 1.25 m in the @RISK analysis shifts the entire Lognormal distribution to the right so that the minimum factor of safety for the assumed conditions is found to be 1.04. The probability of failure for this case is zero. This decrease in bolt spacing would be a prudent practical decision in this case.

It is hoped that this simple example demonstrates that the use of probability theory produces a great deal more information than a simple deterministic factor of safety calculation. Even with the minimal amount of input data which has been used for this case, the shape of the probability distribution curves and the estimated probabilities of failure, for different bolt spacing, can give the designer a feel for the sensitivity of the design and suggest directions in which improvements can be made.

2.6 Problems to which probability cannot be applied

The common factor in the analyses discussed on the previous pages is that a mean factor of safety can be calculated using a relatively simple set of equations. If it is assumed that the distribution of parameters contained in these equations can be described by one of the probability density functions, an analysis of probability of failure can be performed. Unfortunately, this type of analysis is not possible for one of the most important groups of problems in underground excavation engineering, i.e., those problems involving stress driven instability.

Where the rock mass surrounding an underground opening is stressed to the level at which failure initiates, the subsequent behaviour of the rock mass is extremely complex and falls into the category of problems which are classed as 'indeterminate'. In other words, the process of fracture propagation and the deformation of the rock mass surrounding the opening are interactive processes which cannot be represented by a simple set of equations. The study of these problems requires the use of numerical models which follow the process of progressive failure, and the load transfer from failed elements onto unfractured elements until equilibrium is achieved, or until the opening collapses. The introduction of support into such a model further complicates the process, since the capacity and deformational properties of the support influence the behaviour of the rock mass. A model called PHASES, developed specifically for these types of analyses, will be discussed in a later chapter.

A key factor in this analysis of stress driven instability is that there is no clear definition of acceptable stability or of failure. Anyone who has visited a deep level mine will be familiar with the sight of fractured rock surrounding the underground openings and yet these openings are accessible and clearly have not 'failed'. In practical terms, stability is judged to be acceptable when the deformation of the rock mass is controlled and when the support elements are not over-stressed.

While it is not possible to utilise probabilistic techniques, such as the Monte Carlo analysis, directly in the analysis of stress driven instability, it is useful to consider the possible range of input parameters when working with these problems. Hence, when using one of the numerical models to analyse the extent of the failed zone around an opening or the amount of support required to control deformation, it is important to run such a model several times to investigate the influence of variations in applied stresses, rock mass properties and the characteristics of different support systems. With improvements in program efficiency and computer capability, it is becoming feasible to run some of these stress analyses a number of times in a few hours. This means that the user can gain an appreciation for the most probable 'average' conditions which have to be designed for and the possible range of variations which may have to be dealt with in the field.

3 Evaluation of engineering geological data

3.1 Introduction

A rock mass is rarely continuous, homogeneous or isotropic. It is usually intersected by a variety of discontinuities such as faults, joints, bedding planes, and foliation. In addition, there can be a number of different rock types which may have been subjected to varying degrees of alteration or weathering. It is clear that the behaviour of the rock mass, when subjected to the influence of mining excavations, depends on the characteristics of both the rock material and the discontinuities.

A complete engineering geological rock mass description contains details of the rock material and the natural discontinuities. Descriptive indices required to fully characterise the rock mass comprise weathering/alteration, structure, colour, grain size, intact rock material compressive strength and rock type, with details of the discontinuities such as orientation, persistence, spacing, aperture/thickness, infilling, waviness and unevenness for each set. The resulting rock mass can be described by block shape, block size and discontinuity condition. An evaluation of the potential influence of groundwater and the number of joint sets, which will affect the stability of the excavation, completes the description.

Mapping of geological structure is an essential component of the design of underground excavations. Structural planes run through the rock mass and may divide it into discrete blocks of rock, which can fall or slide from the excavation boundary, when they are not adequately supported and when the stress conditions are favourable for structural failure. Data collected from the mapping of these structures are used to determine the orientation of the major joint sets and to assess the potential modes of structural failure.

3.2 Engineering geological data collection

Standardised approaches to the collection of engineering geology data, for civil and mining engineering purposes, have been proposed by the Geological Society of London (Anon., 1977) and by the International Society of Rock Mechanics (ISRM, 1978). It is assumed that the reader is familiar with these techniques or has access to engineering geology data collected by someone who is familiar with these techniques.

The character of the rock mass is comprised of a combination of geological and geometric parameters to which design related or engineering conditions are added during the design process. The main goal in engineering geological data collection is to be able to describe the rock mass as accurately as possible. This will assist in the determination of a rock mass classification as well as providing a means of communication between geologists and engineers working together on a project.

Three examples of typical engineering geological descriptions are:

- Slightly weathered, slightly folded, blocky and schistose, reddish grey, medium grained, strong, Garnet-mica schist with well developed schistosity dipping 45/105. Schistosity is highly persistent, widely spaced, extremely narrow aperture, iron stained, planar, and smooth. Moderate water inflow is expected.
- Slightly weathered, blocky, pale grey, coarse grained, very strong, Granite with three sets of persistent, widely spaced, extremely narrow, iron stained, planar, rough, wet joints.
- Fresh, blocky (medium to large blocks), dark greenish grey, coarse grained, very strong Norite with two sets of persistent, widely to very widely spaced (600 mm), extremely narrow, undulating, rough, dry joints.

Some specific aspects of engineering or structural geology data collection will be discussed in later chapters dealing with the analysis and interpretation of structural data and the estimation of rock mass properties.

3.3 Structural geological terms

Structural geological mapping consists of measuring the orientation of planes (joints, bedding planes or faults) which cut through the rock mass. Other characteristics of these planes, such as the surface roughness, persistence, spacing and weathering may also be measured and incorporated into rock mass classification schemes discussed in the next chapter. The orientation and inclination of any structural plane are defined by two measurements which can be expressed as either dip and dip direction or strike and dip. Dip and dip direction are more useful for engineering purposes and for the processing of structural geology by computer, while dip and strike are the terms which are generally used by geologists working in the field. The definitions of these terms are illustrated in Figure 3.1.

One of the easiest ways in which to visualise the definition of the terms dip and dip direction is to imagine a ball rolling down a plane. The ball will roll down the line of maximum inclination of the plane and this line defines both the dip and the dip direction of the plane. The vertical angle of the line of maximum inclination, measured from a horizontal plane, is defined as the dip. The orientation of the horizontal projection of the line of maximum inclination, measured clockwise from north, is the dip direction. The strike of the plane is the direction of the line of intersection of the plane and a horizontal surface.

By convention, the dip and dip direction measurements are generally written as 35/120, where the two digit number refers to the dip and the three digit number refers to the dip direction. The corresponding strike and dip values are generally written as 030/35SE or 030/35, using the right hand rule[1].

---

[1] The right hand rule is: with your right hand palm down and your fingers pointing down dip, your thumb indicates the direction of the strike.

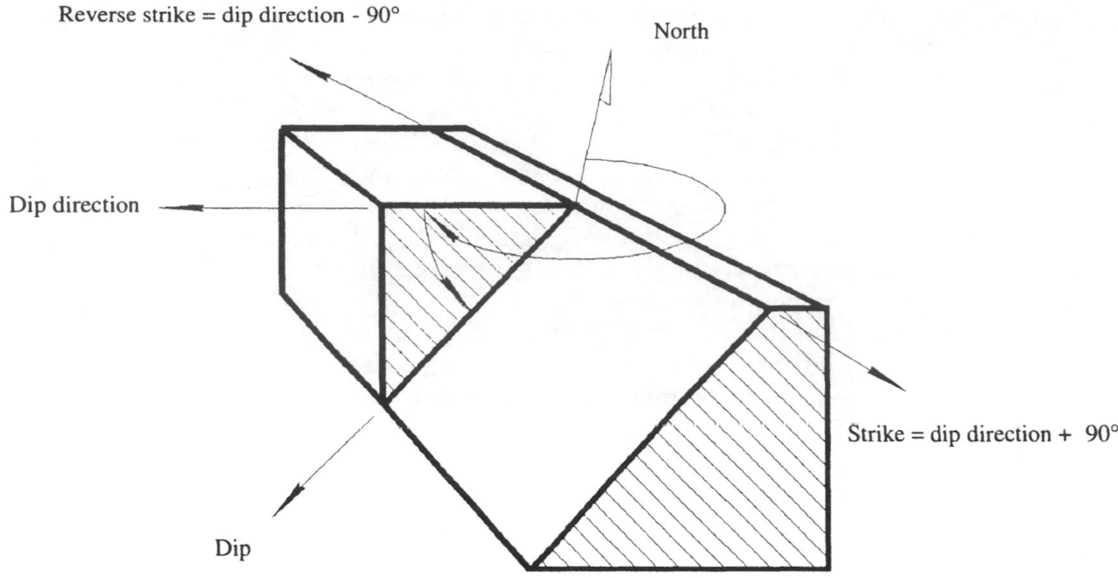

Figure 3.1: Definition of strike, dip and dip direction.

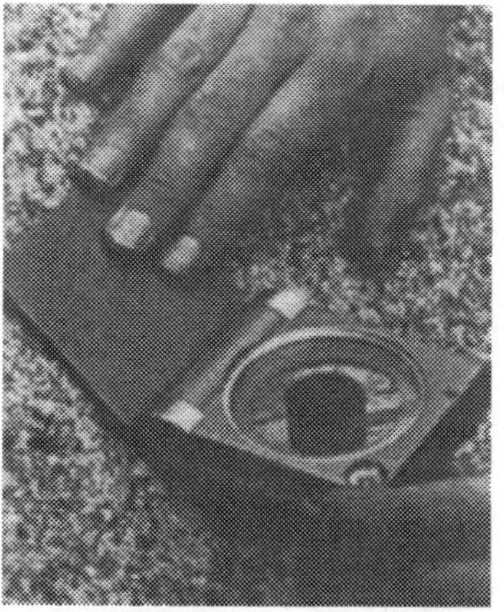

Clar compass manufactured by F.W. Breithaupt & Sohn, Kassel, Germany.

Tectronic 4000 compass for electronic measurement and storage of structural data. Manufactured by F.W. Breithaupt & Sohn, Kassel, Germany.

3.4 Structural geological data collection

Many types of compasses and clinorules are available for measuring the orientation of planes. Some of these are more convenient than others for use in underground openings. The advantages and disadvantages of some of these instruments are discussed in Hoek and Bray (1981). The choice of instrument is generally a matter of personal preference or budget constraints and it is advisable to discuss this choice with an experienced engineering geologist before purchasing a compass.

Geological data collection should be methodical to ensure that all relevant data are collected. Therefore, it is wise to establish scanlines, or 'windows' on the rock face, where structural measurements will be made. All significant features, which cross these lines or are contained in the windows, are recorded in the geological mapping. In this context *significant* generally means that the trace of the feature should be clearly visible to the naked eye and should be more than a metre long. The scanlines selected should be oriented in as many directions as possible to provide maximum coverage of the joint sets.

Whenever possible, at least 100 measurements of dip and dip direction (or dip and strike) should be made in each structural domain, which is a block of ground considered to have uniform properties. Some bias will always be present in the geological data set. This bias arises from the fact that the features oriented perpendicular to the traverse will be closest to the true spacing, while features oriented sub-parallel to the surface being mapped will appear to be more widely spaced than they actually are, and fewer measurements of the latter will be made. A correction for sampling bias can be incorporated into the analysis of the structural data, as it is done in the microcomputer program DIPS[2].

3.5 Structural geological data presentation

The presentation of the structural geological data collected at a site is most conveniently done using the spherical projection technique, in which a plane in three-dimensional space is represented by a great circle on a two-dimensional projection. This is exactly the same technique used by map makers to represent the spherical earth on a two-dimensional map.

An extremely important point to note is that planes are assumed to be ubiquitous, i.e., they can occur anywhere in space. This allows us to arrange them in such a way that they all pass through the centre of the reference sphere. The assumption of ubiquity will become increasingly important through the balance of the discussion.

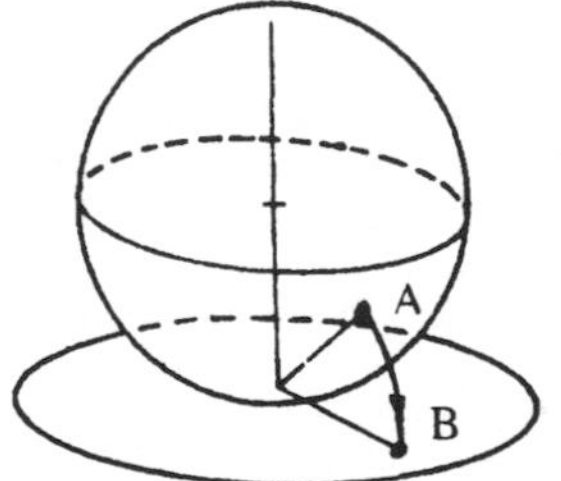

Equal area projection

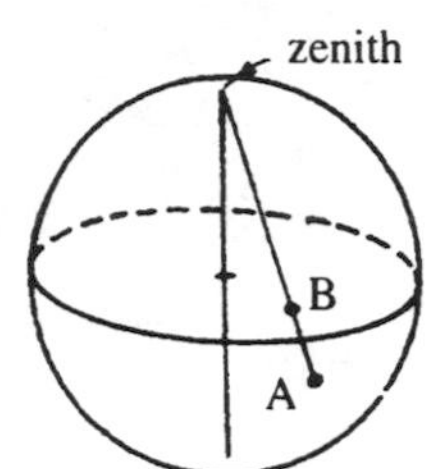

Equal angle projection

A single plane oriented in three dimensional space is shown in Figure 3.2. The intersection of the plane with the reference sphere, shown in this figure as a shaded part ellipse, defines a great circle when projected in two-dimensional space. A pole is defined at the point where a line, drawn through the centre of the sphere perpendicular to the plane, intersects the sphere.

The projections of the circle and poles to a two-dimensional horizontal plane are constructed following one of two conventions: the equal area or the equal angle projection.

In the equal area method the bottom of the sphere rests on the projection plane. The point A on the sphere is projected down to the plane by swinging this point in an arc about the contact between the sphere and the plane, giving point B.

In the case of the equal angle projection, a line is drawn from the centre of the top of the sphere (the zenith) to the point A on the sphere. The intersection of this line, with a horizontal plane through the centre of the sphere, defines the projection point B.

Note that, in both cases illustrated in the margin sketch, the point A lies on the lower hemisphere and these projections are referred to as lower hemisphere projections. Lower hemisphere projections are used throughout this book.

When used for structural data analysis, as discussed below, the two projection methods produce practically identical results. When the analyses are carried out manually, as described in Hoek and Bray (1981) or Hoek and Brown (1980a), each method has advantages and disadvantages, depending upon the particular type of analysis being performed.

When the analyses are carried out by means of a computer program, such as DIPS, there is no difference between the mean pole calculations made by the two methods, so the choice of which projection to use becomes a matter of personal preference. Whether

[2] This program is available from Rock Engineering Group, 12 Selwood Avenue, Toronto, Ontario, Canada M4E 1B2, Fax 1 416 698 0908, Phone 1 416 698 8217. (See order form at the end of this book).

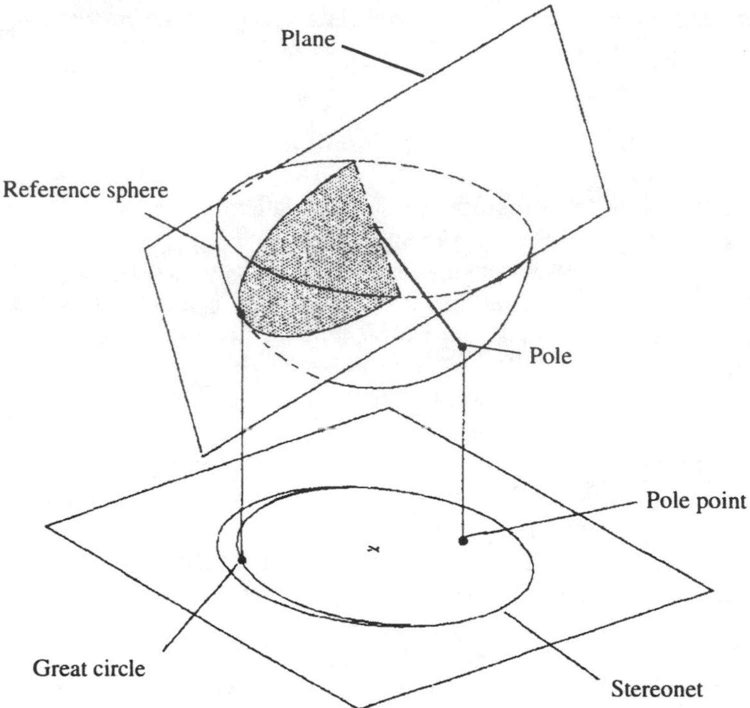

Figure 3.2: Definition of a great circle and pole.

manual or computer methods are used, it is essential that the two projections should never be mixed. Decide upon one or the other and use that projection for all data presentation and analysis on a project.

3.6 Geological data analysis

A set of dip and dip direction measurements is most conveniently plotted as poles on a stereonet (the generic name used to describe the diagram produced when using one of the spherical projection techniques described above). A typical plot of 61 poles is given in Figure 3.3. Note that different symbols are used to indicate locations on the stereonet where two or more poles are coincident.

The poles plotted in Figure 3.3 were measured in an exploration adit in gneiss with a few well developed joint sets. It is difficult to discern the different sets in the plot given in Figure 3.3, because of the scatter in the poles as a result of local variations in the dip and dip direction of the individual features. Consequently, in order to establish the average orientation of each family of significant discontinuities, the poles are contoured to produce the diagram in Figure 3.4.

A number of manual contouring techniques are discussed in Hoek and Bray (1981) and Hoek and Brown (1980a) and the choice of which technique to use is a matter of individual preference. The contour plot given in Figure 3.4 was produced using the program DIPS.

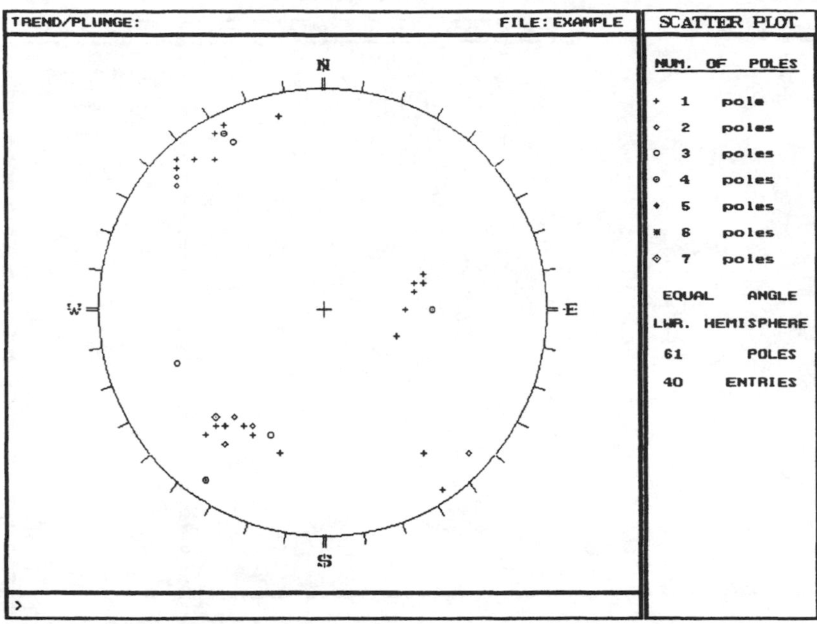

Figure 3.3: Scatter plot of 61 poles on an equal area lower hemisphere projection.

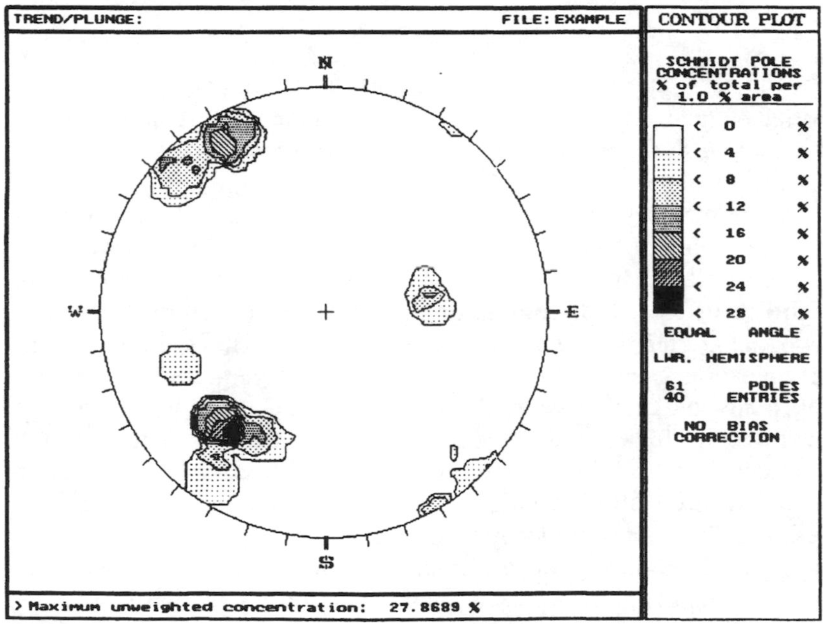

Figure 3.4: Pole density contour plot for the scatter plot illustrated in Figure 3.3.

Note that, although all structural features can be plotted on a stereonet, the inclusion of a single pole, representing a fault or a major shear zone, in the data being contoured could result in this feature being lost in the counting process which does not assign weights to individual poles. Consequently, when a fault or major shear zone is present in the rock mass being considered, it is advisable to use a different symbol to plot the pole representing this feature. This pole is then clearly identified as a major feature requiring

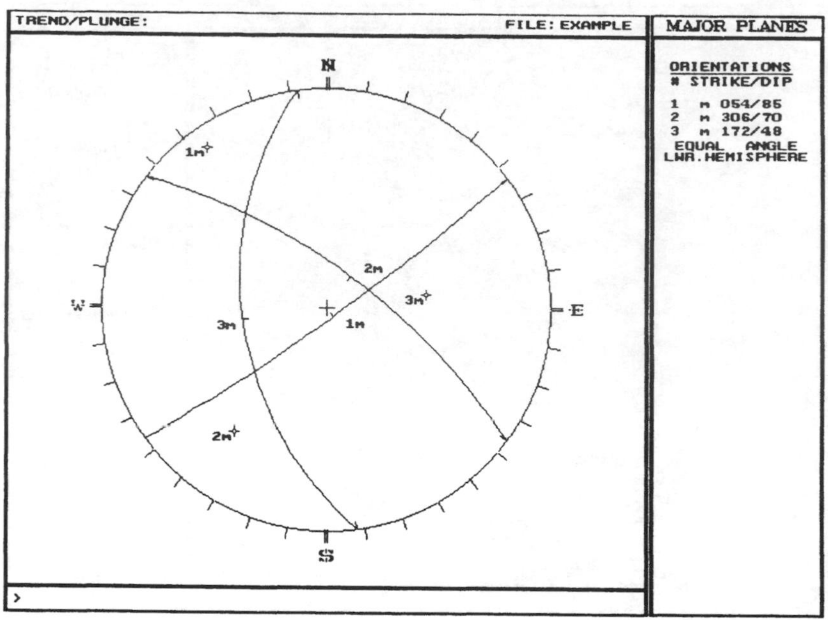

Figure 3.5: Poles and corresponding great circles for the average dip and dip direction of 3 discontinuity sets represented by the contour plot shown in Figure 3.4.

special consideration. This is particularly important where the data is collected and analysed by one individual and then passed on to someone else for incorporation into a stability analysis or for inclusion in a support design.

Once the contours have been plotted, the average dip and dip direction values for each discontinuity set are found by locating the highest pole density in each contour cluster. Where the contours are tightly clustered, indicating strongly developed planar features such as bedding planes in undeformed sedimentary rocks, these high density locations are easy to determine by eye. Where there is more scatter in the pole plot, as would be the case for rock masses, which have been locally folded and faulted, it is more difficult to determine the average strike and dip of each set by visual inspection. In such cases, a statistical counting technique is applied to each contour cluster in order to determine the location of the highest contour density. The program DIPS allows for different counting procedures to assist in determining the point representing the maximum pole density.

Application of these contouring procedures gives the great circle plot in Figure 3.5. This plot defines the average dips and dip directions of significant bedding planes, joints and other structural features in a rock mass. This information can then be used in the structural stability analyses and support design procedures described later in this book.

4 Rock mass classification

4.1 Introduction

During the feasibility and preliminary design stages of a project, when very little detailed information on the rock mass and its stress and hydrologic characteristics is available, the use of a rock mass classification scheme can be of considerable benefit. At its simplest, this may involve using the classification scheme as a check-list to ensure that all relevant information has been considered. At the other end of the spectrum, one or more rock mass classification schemes can be used to build up a picture of the composition and characteristics of a rock mass to provide initial estimates of support requirements, and to provide estimates of the strength and deformation properties of the rock mass.

It is important to understand that the use of a rock mass classification scheme does not (and cannot) replace some of the more elaborate design procedures. However, the use of these design procedures requires access to relatively detailed information on in situ stresses, rock mass properties and planned excavation sequence, none of which may be available at an early stage in the project. As this information becomes available, the use of the rock mass classification schemes should be updated and used in conjunction with site specific analyses.

4.2 Engineering rock mass classification

Rock mass classification schemes have been developing for over 100 years since Ritter (1879) attempted to formalise an empirical approach to tunnel design, in particular for determining support requirements. While the classification schemes are appropriate for their original application, especially if used within the bounds of the case histories from which they were developed, considerable caution must be exercised in applying rock mass classifications to other rock engineering problems.

Summaries of some important classification systems are presented in this chapter, and although every attempt has been made to present all of the pertinent data from the original texts, there are numerous notes and comments which cannot be included. The interested reader should make every effort to read the cited references for a full appreciation of the use, applicability and limitations of each system.

Most of the multi-parameter classification schemes (Wickham et al., 1972, Bieniawski, 1973, 1989, and Barton et al., 1974) were developed from civil engineering case histories in which all of the components of the engineering geological character of the rock mass were included. In underground hard rock mining, however, especially at deep levels, rock mass weathering and the influence of water usually are not significant and may be ignored. Different classification systems place different emphases on the various parameters, and it is recommended that at least two methods be used at any site during the early stages of a project.

4.2.1 *Terzaghi's rock mass classification*

The earliest reference to the use of rock mass classification for the design of tunnel support is in a paper by Terzaghi (1946) in which the rock loads, carried by steel sets, are estimated on the basis of a descriptive classification. While no useful purpose would be served by including details of Terzaghi's classification in this discussion on the design of support for underground hard rock mines, it is interesting to examine the rock mass descriptions included in his original paper, because he draws attention to those characteristics that dominate rock mass behaviour, particularly in situations where gravity constitutes the dominant driving force. The clear and concise definitions and the practical comments included in these descriptions are good examples of the type of engineering geology information, which is most useful for engineering design.

Terzaghi's descriptions (quoted directly from his paper) are:

- *Intact* rock contains neither joints nor hair cracks. Hence, if it breaks, it breaks across sound rock. On account of the injury to the rock due to blasting, spalls may drop off the roof several hours or days after blasting. This is known as a *spalling* condition. Hard, intact rock may also be encountered in the *popping* condition involving the spontaneous and violent detachment of rock slabs from the sides or roof.
- *Stratified* rock consists of individual strata with little or no resistance against separation along the boundaries between the strata. The strata may or may not be weakened by transverse joints. In such rock the spalling condition is quite common.
- *Moderately jointed* rock contains joints and hair cracks, but the blocks between joints are locally grown together or so intimately interlocked that vertical walls do not require lateral support. In rocks of this type, both spalling and popping conditions may be encountered.
- *Blocky and seamy* rock consists of chemically intact or almost intact rock fragments which are entirely separated from each other and imperfectly interlocked. In such rock, vertical walls may require lateral support.
- *Crushed* but chemically intact rock has the character of crusher run. If most or all of the fragments are as small as fine sand grains and no recementation has taken place, crushed rock below the water table exhibits the properties of a water-bearing sand.
- *Squeezing* rock slowly advances into the tunnel without perceptible volume increase. A prerequisite for squeeze is a high percentage of microscopic and sub-microscopic particles of micaceous minerals or clay minerals with a low swelling capacity.
- *Swelling* rock advances into the tunnel chiefly on account of expansion. The capacity to swell seems to be limited to those rocks that contain clay minerals such as montmorillonite, with a high swelling capacity.

4.2.2 *Classifications involving stand-up time*

Lauffer (1958) proposed that the stand-up time for an unsupported span is related to the quality of the rock mass in which the span is excavated. In a tunnel, the unsupported span is defined as the span of the tunnel or the distance between the face and the nearest support, if this is greater than the tunnel span. Lauffer's original classification has since been modified by a number of authors, notably Pacher et al. (1974), and now forms part of the general tunnelling approach known as the New Austrian Tunnelling Method.

The significance of the stand-up time concept is that an increase in the span of the tunnel leads to a significant reduction in the time available for the installation of support. For example, a small pilot tunnel may be successfully constructed with minimal support, while a larger span tunnel in the same rock mass may not be stable without the immediate installation of substantial support.

The New Austrian Tunnelling Method includes a number of techniques for safe tunnelling in rock conditions in which the stand-up time is limited before failure occurs. These techniques include the use of smaller headings and benching or the use of multiple drifts to form a reinforced ring inside which the bulk of the tunnel can be excavated. These techniques are applicable in soft rocks such as shales, phyllites and mudstones in which the squeezing and swelling problems, described by Terzaghi (see previous section), are likely to occur. The techniques are also applicable when tunnelling in excessively broken rock, but great care should be taken in attempting to apply these techniques to excavations in hard rocks in which different failure mechanisms occur.

In designing support for hard rock excavations it is prudent to assume that the stability of the rock mass surrounding the excavation is not time-dependent. Hence, if a structurally defined wedge is exposed in the roof of an excavation, it will fall as soon as the rock supporting it is removed. This can occur at the time of the blast or during the subsequent scaling operation. If it is required to keep such a wedge in place, or to enhance the margin of safety, it is essential that the support be installed as early as possible, preferably before the rock supporting the full wedge is removed. On the other hand, in a highly stressed rock, failure will generally be induced by some change in the stress field surrounding the excavation. The failure may occur gradually and manifest itself as spalling or slabbing or it may occur suddenly in the form of a rock burst. In either case, the support design must take into account the change in the stress field rather than the 'stand-up' time of the excavation.

4.2.3 *Rock quality designation index (RQD)*

The Rock Quality Designation index (*RQD*) was developed by Deere (Deere et al., 1967) to provide a quantitative estimate of rock mass quality from drill core logs. *RQD* is defined as the percentage of intact core pieces longer than 100 mm (4 inches) in the total length of core. The core should be at least NX size (54.7 mm or 2.15 inches in diameter) and should be drilled with a double-tube core barrel. The correct procedures for measurement of the length of core pieces and the calculation of *RQD* are summarised in Figure 4.1.

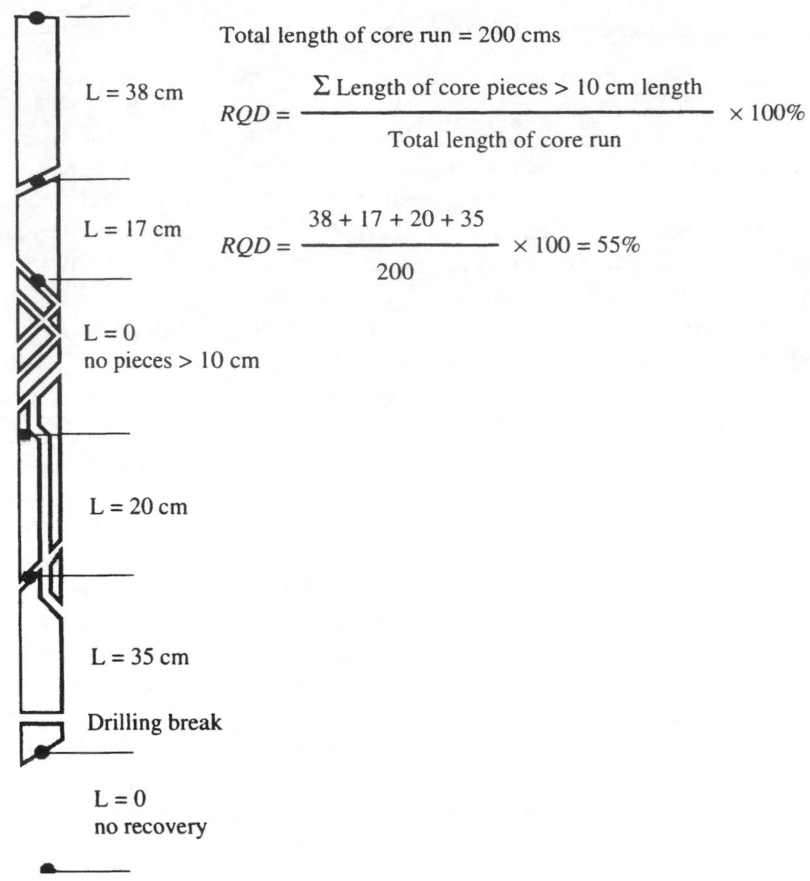

Figure 4.1: Procedure for measurement and calculation of *RQD* (After Deere, 1989).

Palmström (1982) suggested that, when no core is available but discontinuity traces are visible in surface exposures or exploration adits, the *RQD* may be estimated from the number of discontinuities per unit volume. The suggested relationship for clay-free rock masses is:

$$RQD = 115\text{-}3.3\ J_v \tag{4.1}$$

where J_v is the sum of the number of joints per unit length for all joint (discontinuity) sets known as the volumetric joint count.

RQD is a directionally dependent parameter and its value may change significantly, depending upon the borehole orientation. The use of the volumetric joint count can be quite useful in reducing this directional dependence.

RQD is intended to represent the rock mass quality in situ. When using diamond drill core, care must be taken to ensure that fractures, which have been caused by handling or the drilling process, are identified and ignored when determining the value of *RQD*. When using Palmström's relationship for exposure mapping, blast induced fractures should not be included when estimating J_v.

Deere's *RQD* has been widely used, particularly in North America, for the past 25 years. Cording and Deere (1972), Merritt (1972) and Deere and Deere (1988) have attempted to relate *RQD* to Terzaghi's rock load factors and to rockbolt requirements in tunnels. In the context of this discussion, the most important use of *RQD* is as a component of the *RMR* and *Q* rock mass classifications covered later in this chapter.

4.2.4 *Rock Structure Rating (RSR)*

Wickham et al. (1972) described a quantitative method for describing the quality of a rock mass and for selecting appropriate support on the basis of their Rock Structure Rating (*RSR*) classification. Most of the case histories, used in the development of this system, were for relatively small tunnels supported by means of steel sets, although historically this system was the first to make reference to shotcrete support. In spite of this limitation, it is worth examining the *RSR* system in some detail since it demonstrates the logic involved in developing a quasi-quantitative rock mass classification system and the utilisation of the resulting index for support estimation.

The significance of the *RSR* system, in the context of this discussion, is that it introduced the concept of rating each of the components listed below to arrive at a numerical value of $RSR = A + B + C$.

1. *Parameter A, Geology:* General appraisal of geological structure on the basis of:
   a. Rock type origin (igneous, metamorphic, sedimentary).
   b. Rock hardness (hard, medium, soft, decomposed).
   c. Geologic struc ure (massive, slightly faulted/folded, moderately faulted/folded, intensely faulted/folded).
2. *Parameter B, Geometry*: Effect of discontinuity pattern with respect to the direction of the tunnel drive on the basis of:
   a. Joint spacing.
   b. Joint orientation (strike and dip).
   c. Direction of tunnel drive.
3. *Parameter C*: Effect of groundwater inflow and joint condition on the basis of:
   a. Overall rock mass quality on the basis of A and B combined.
   b. Joint condition (good, fair, poor).
   c. Amount of water inflow (in gallons per minute per 1000 feet of tunnel).

Note that the *RSR* classification used Imperial units and that these units have been retained in this discussion.

Three tables from Wickham et al.'s 1972 paper are reproduced in Tables 4.1, 4.2 and 4.3. These tables can be used to evaluate the rating of each of these parameters to arrive at the *RSR* value (maximum $RSR = 100$).

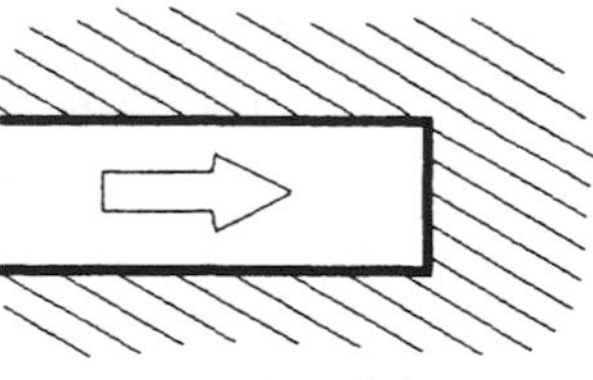

Drive with dip

For example, a hard metamorphic rock which is slightly folded or faulted has a rating of $A = 22$ (from Table 4.1). The rock mass is moderately jointed, with joints striking perpendicular to the tunnel axis which is being driven east-west, and dipping at between 20° and 50°. Table 4.2 gives the rating for $B = 24$ for driving with dip (defined in the margin sketch). The value of $A + B = 46$ and this means that, for joints of fair condition (slightly weathered and altered) and a moderate water inflow of between 200 and 1,000 gallons per minute, Table 4.3 gives the rating for $C = 16$. Hence, the final value of the rock structure rating $RSR = A + B + C = 62$.

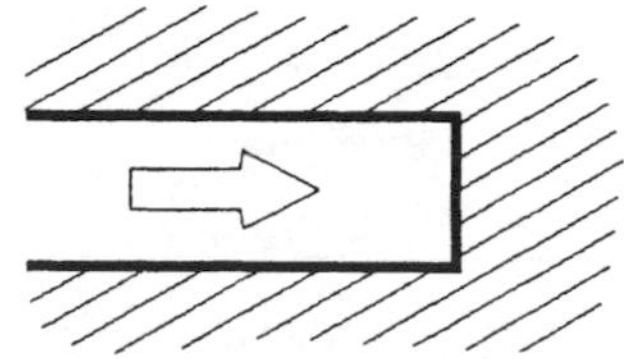

Drive against dip

Table 4.1: Rock Structure Rating: Parameter *A*: General area geology.

	Basic Rock Type				Geological Structure			
	Hard	Medium	Soft	Decomposed		Slightly	Moderately	Intensively
Igneous	1	2	3	4		Folded or	Folded or	Folded or
Metamorphi	1	2	3	4	Massive	Faulted	Faulted	Faulted
Sedimentary	2	3	4	4				
Type 1					30	22	15	9
Type 2					27	20	13	8
Type 3					24	18	12	7
Type 4					19	15	10	6

Table 4.2: Rock Structure Rating: Parameter *B*: Joint pattern, direction of drive.

	Strike ⊥ to Axis					Strike ǁ to Axis		
	Direction of Drive					Direction of Drive		
	Both	With Dip		Against Dip		Either direction		
	Dip of Prominent Joints [a]					Dip of Prominent Joints		
Average joint spacing	Flat	Dipping	Vertical	Dipping	Vertical	Flat	Dipping	Vertical
1. Very closely jointed, < 2	9	11	13	10	12	9	9	7
2. Closely jointed, 2-6 in	13	16	19	15	17	14	14	11
3. Moderately jointed, 6-12	23	24	28	19	22	23	23	19
4. Moderate to blocky, 1-2 ft	30	32	36	25	28	30	28	24
5. Blocky to massive, 2-4 ft	36	38	40	33	35	36	24	28
6. Massive, > 4 ft	40	43	45	37	40	40	38	34

Table 4.3: Rock Structure Rating: Parameter *C*: Groundwater, joint condition.

	Sum of Parameters *A* + *B*					
	13-44			45-75		
Anticipated water inflow	Joint Condition [b]					
gpm/1000 ft of tunnel	Good	Fair	Poor	Good	Fair	Poor
None	22	18	12	25	22	18
Slight, < 200 gpm	19	15	9	23	19	14
Moderate, 200-1000 gpm	15	22	7	21	16	12
Heavy, > 1000 gpm	10	8	6	18	14	10

[a] Dip: flat: 0-20°; dipping: 20-50°; and vertical: 50-90°.
[b] Joint condition: good = tight or cemented; fair = slightly weathered or altered; poor = severely weathered, altered or open.

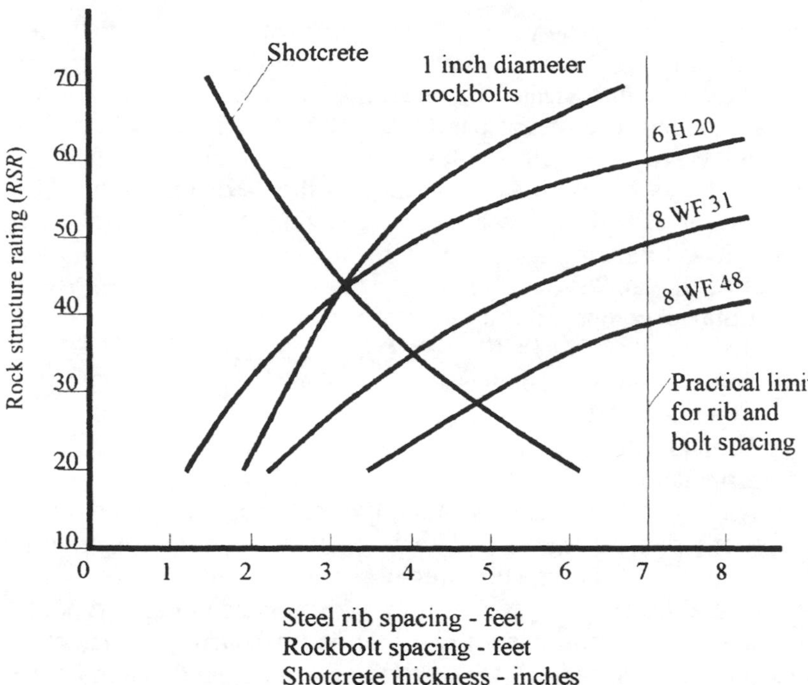

Figure 4.2: *RSR* support estimates for a 24 ft. (7.3 m) diameter circular tunnel. Note that rockbolts and shotcrete are generally used together. (After Wickham et al., 1972).

A typical set of prediction curves for a 24 foot diameter tunnel are given in Figure 4.2 which shows that, for the *RSR* value of 62 derived above, the predicted support would be 2 inches of shotcrete and 1 inch diameter rockbolts spaced at 5 foot centres. As indicated in the figure, steel sets would be spaced at more than 7 feet apart and would not be considered a practical solution for the support of this tunnel.

For the same size tunnel in a rock mass with *RSR* = 30, the support could be provided by 8 WF 31 steel sets (8 inch deep wide flange I section weighing 31 lb per foot) spaced 3 feet apart, or by 5 inches of shotcrete and 1 inch diameter rockbolts spaced at 2.5 feet centres. In this case it is probable that the steel set solution would be cheaper and more effective than the use of rockbolts and shotcrete.

The reader should be aware that these estimates are very crude, particularly for rockbolts and shotcrete, since they are based upon a relatively small number of case histories and very simplistic theoretical arguments. Consequently, they should be applied with great caution.

Although the *RSR* classification system is not widely used, particularly in mining, Wickham et al.'s work played a significant role in the development of the classification schemes discussed in the remaining sections of this chapter.

4.3 Geomechanics Classification

Bieniawski (1976) published the details of a rock mass classification called the Geomechanics Classification or the Rock Mass Rating

(*RMR*) system. Over the years, this system has been successively refined as more case records have been examined and the reader should be aware that Bieniawski has made significant changes in the ratings assigned to different parameters. The discussion which follows is based upon the 1989 version of the classification (Bieniawski, 1989). Both this version and the 1976 version will be used in Chapter 8 which deals with estimating the strength of rock masses. The following six parameters are used to classify a rock mass using the *RMR* system:

1. Uniaxial compressive strength of rock material.
2. Rock Quality Designation (*RQD*).
3. Spacing of discontinuities.
4. Condition of discontinuities.
5. Groundwater conditions.
6. Orientation of discontinuities.

In applying this classification system, the rock mass is divided into a number of structural regions and each region is classified separately. The boundaries of the structural regions usually coincide with a major structural feature such as a fault or with a change in rock type. In some cases, significant changes in discontinuity spacing or characteristics, within the same rock type, may necessitate the division of the rock mass into a number of small structural regions or domains.

The Rock Mass Rating system is presented in Table 4.4, giving the ratings for each of the six parameters listed above. These ratings are summed to give a value of *RMR*. The following example illustrates the use of these tables to arrive at an *RMR* value.

A tunnel is to be driven through a slightly weathered granite with a dominant joint set dipping at 60° against the direction of the drive. Index testing and logging of diamond drilled core give typical Point-load strength index values of 8 MPa and average *RQD* values of 70%. The joints, which are slightly rough and slightly weathered with a separation of < 1 mm, are spaced at 300 mm. Tunnelling conditions are anticipated to be wet.

The *RMR* value is determined as follows:

Table	Item	Value	Rating
4.1: A.1	Point load index	8 MPa	12
4.1: A.2	*RQD*	70%	13
4.1: A.3	Spacing of discontinuities	300 mm	10
4.1: E.4	Condition of discontinuities	Note 1	22
4.1: A.5	Groundwater	Wet	7
4.1: B	Adjustment for joint orientation	Note 2	-5
		Total	59

Note 1. For slightly rough and altered discontinuity surfaces with a separation of < 1 mm, Table 4.4.A.4 gives a rating of 25. When more detailed information is available, Table 4.4.E can be used to obtain a more refined rating. Hence, in this case, the rating is the sum of: 4 (1-3 m discontinuity length), 4 (separation 0.1-1.0 mm), 3 (slightly rough), 6 (no infilling) and 5 (slightly weathered) = 22.

Note 2. Table 4.4.F gives a description of 'Fair' for the conditions assumed where the tunnel is to be driven against the dip of a set of joints dipping at 60°. Using this description for 'Tunnels and Mines' in Table 4.4.B gives an adjustment rating of -5.

Table 4.4: Rock Mass Rating System (After Bieniawski, 1989).

A. CLASSIFICATION PARAMETERS AND THEIR RATINGS

	Parameter		Range of values						
1	Strength of intact rock material	Point-load strength index	>10 MPa	4-10 MPa	2-4 MPa	1-2 MPa	For this low range - uniaxial compressive test is preferred		
		Uniaxial comp. strength	>250 MPa	100-250 MPa	50-100 MPa	25-50 MPa	5-25 MPa	1-5 MPa	< 1 MPa
	Rating		15	12	7	4	2	1	0
2	Drill core Quality *RQD*		90%-100%	75%-90%	50%-75%	25%-50%	< 25%		
	Rating		20	17	13	8	3		
3	Spacing of discontinuities		> 2 m	0.6-2 . m	200-600 mm	60-200 mm	< 60 mm		
	Rating		20	15	10	8	5		
4	Condition of discontinuities (See E)		Very rough surfaces Not continuous No separation Unweathered wall rock	Slightly rough surfaces Separation < 1 mm Slightly weathered walls	Slightly rough surfaces Separation < 1 mm Highly weathered walls	Slickensided surfaces or Gouge < 5 mm thick or Separation 1-5 mm Continuous	Soft gouge >5 mm thick or Separation > 5 mm Continuous		
	Rating		30	25	20	10	0		
5	Ground water	Inflow per 10 m tunnel length (l/m)	None	< 10	10-25	25-125	> 125		
		(Joint water press)/ (Major principal σ)	0	< 0.1	0.1,-0.2	0.2-0.5	> 0.5		
		General conditions	Completely dry	Damp	Wet	Dripping	Flowing		
	Rating		15	10	7	4	0		

B. RATING ADJUSTMENT FOR DISCONTINUITY ORIENTATIONS (See F)

Strike and dip orientations		Very favourable	Favourable	Fair	Unfavourable	Very Unfavourable
Ratings	Tunnels & mines	0	-2	-5	-10	-12
	Foundations	0	-2	-7	-15	-25
	Slopes	0	-5	-25	-50	

C. ROCK MASS CLASSES DETERMINED FROM TOTAL RATINGS

Rating	100 ← 81	80 ← 61	60 ← 41	40 ← 21	< 21
Class number	I	II	III	IV	V
Description	Very good rock	Good rock	Fair rock	Poor rock	Very poor rock

D. MEANING OF ROCK CLASSES

Class number	I	II	III	IV	V
Average stand-up time	20 yrs for 15 m span	1 year for 10 m span	1 week for 5 m span	10 hrs for 2.5 m span	30 min for 1 m span
Cohesion of rock mass (kPa)	> 400	300-400	200-300	100-200	< 100
Friction angle of rock mass (deg)	> 45	35-45	25-35	15-25	< 15

E. GUIDELINES FOR CLASSIFICATION OF DISCONTINUITY conditions

Discontinuity length (persistence) Rating	< 1 m 6	1-3 m 4	3-10 m 2	10-20 m 1	> 20 m 0
Separation (aperture) Rating	None 6	< 0.1 mm 5	0.1-1.0 mm 4	1-5 mm 1	> 5 mm 0
Roughness Rating	Very rough 6	Rough 5	Slightly rough 3	Smooth 1	Slickensided 0
Infilling (gouge) Rating	None 6	Hard filling < 5 mm 4	Hard filling > 5 mm 2	Soft filling < 5 mm 2	Soft filling > 5 mm 0
Weathering Ratings	Unweathered 6	Slightly weathered 5	Moderately weathered 3	Highly weathered 1	Decomposed 0

F. EFFECT OF DISCONTINUITY STRIKE AND DIP ORIENTATION IN TUNNELLING**

Strike perpendicular to tunnel axis		Strike parallel to tunnel axis	
Drive with dip-Dip 45-90°	Drive with dip-Dip 20-45°	Dip 45-90°	Dip 20-45°
Very favourable	Favourable	Very favourable	Fair
Drive against dip-Dip 45-90°	Drive against dip-Dip 20-45°	Dip 0-20-Irrespective of strike°	
Fair	Unfavourable	Fair	

*Some conditions are mutually exclusive. For example, if infilling is present, the roughness of the surface will be overshadowed by the influence of the gouge. In such cases use A.4 directly.

**Modified after Wickham et al. (1972).

Bieniawski (1989) published a set of guidelines for the selection of support in tunnels in rock for which the value of *RMR* has been determined. These guidelines are reproduced in Table 4.5. Note that these guidelines have been published for a 10 m span horseshoe shaped tunnel, constructed using drill and blast methods, in a rock mass subjected to a vertical stress < 25 MPa (equivalent to a depth below surface of <900 m).

For the case considered earlier, with *RMR* = 59, Table 4.5 suggests that a tunnel could be excavated by top heading and bench, with a 1.5 to 3 m advance in the top heading. Support should be installed after each blast and the support should be placed at a maximum distance of 10 m from the face. Systematic rock bolting, using 4 m long 20 mm diameter fully grouted bolts spaced at 1.5 to 2 m in the crown and walls, is recommended. Wire mesh, with 50 to 100 mm of shotcrete for the crown and 30 mm of shotcrete for the walls, is recommended.

The value of *RMR* of 59 indicates that the rock mass is on the boundary between the 'Fair rock' and 'Good rock' categories. In the initial stages of design and construction, it is advisable to utilise the support suggested for fair rock. If the construction is progressing well with no stability problems, and the support is performing very well, then it should be possible to gradually reduce the support requirements to those indicated for a good rock mass. In addition, if the excavation is required to be stable for a short amount of time, then it is advisable to try the less expensive and extensive support suggested for good rock. However, if the rock mass surrounding the excavation is expected to undergo large mining induced stress

Table 4.5: Guidelines for excavation and support of 10 m span rock tunnels in accordance with the *RMR* system (After Bieniawski, 1989).

Rock mass class	Excavation	Rock bolts (20 mm diameter, fully grouted)	Shotcrete	Steel sets
I – Very good rock *RMR*: 81-100	Full face, 3 m advance	Generally no support required except spot bolting		
II – Good rock *RMR*: 61-80	Full face , 1-1.5 m advance. Complete support 20 m from face	Locally, bolts in crown 3 m long, spaced 2.5 m with occasional wire mesh	50 mm in crown where required	None
III – Fair rock *RMR*: 41-60	Top heading and bench 1.5-3 m advance in top heading. Commence support after each blast. Complete support 10 m from face	Systematic bolts 4 m long, spaced 1.5-2 m in crown and walls with wire mesh in crown	50-100 mm in crown and 30 mm in sides	None
IV – Poor rock *RMR*: 21-40	Top heading and bench 1.0-1.5 m advance in top heading. Install support concurrently with excavation, 10 m from face	Systematic bolts 4-5 m long, spaced 1-1.5 m in crown and walls with wire mesh	100-150 mm in crown and 100 mm in sides	Light to medium ribs spaced 1.5 m where required
V – Very poor rock *RMR*: < 20	Multiple drifts 0.5-1.5 m advance in top heading. Install support concurrently with excavation. Shotcrete as soon as possible after blasting	Systematic bolts 5-6 m long, spaced 1-1.5 m in crown and walls with wire mesh. Bolt invert	150-200 mm in crown, 150 mm in sides, and 50 mm on face	Medium to heavy ribs spaced 0.75 m with steel lagging and forepoling if required. Close invert

changes, then more substantial support appropriate for fair rock should be installed. This example indicates that a great deal of judgement is needed in the application of rock mass classification to support design.

It should be noted that Table 4.5 has not had a major revision since 1973. In many mining and civil engineering applications, steel fibre reinforced shotcrete may be considered in place of wire mesh and shotcrete.

4.4 Modifications to *RMR* for mining

Bieniawski's Rock Mass Rating (*RMR*) system was originally based upon case histories drawn largely from civil engineering. Consequently, the mining industry tended to regard the classification as somewhat conservative and several modifications have been proposed in order to make the classification more relevant to mining applications.

A full discussion of all of these modifications would exceed the scope of this volume and the interested reader is referred to the comprehensive summary compiled by Bieniawski (1989).

Laubscher (1977, 1984), Laubscher and Taylor (1976) and Laubscher and Page (1990) have described a Modified Rock Mass Rating system for mining. This *MRMR* system takes the basic *RMR* value, as defined by Bieniawski, and adjusts it to account for in situ and induced stresses, stress changes and the effects of blasting and weathering. A set of support recommendations is associated with the resulting *MRMR* value. In using Laubscher's *MRMR* system it should be borne in mind that many of the case histories upon which it is based are derived from caving operations. Originally, block caving in asbestos mines in Africa formed the basis for the modifications but, subsequently, other case histories from around the world have been added to the database.

Cummings et al. (1982) and Kendorski et al. (1983) have also modified Bieniawski's *RMR* classification to produce the *MBR* (modified basic *RMR*) system for mining. This system was developed for block caving operations in the USA. It involves the use of different ratings for the original parameters used to determine the value of *RMR* and the subsequent adjustment of the resulting *MBR* value to allow for blast damage, induced stresses, structural features, distance from the cave front and size of the caving block. Support recommendations are presented for isolated or development drifts as well as for the final support of intersections and drifts.

4.5 Rock Tunnelling Quality Index, Q

On the basis of an evaluation of a large number of case histories of underground excavations, Barton et al. (1974) of the Norwegian Geotechnical Institute proposed a Tunnelling Quality Index (Q) for the determination of rock mass characteristics and tunnel support requirements. The numerical value of the index Q varies on a logarithmic scale from 0.001 to a maximum of 1,000 and is defined by:

$$Q = \frac{RQD}{J_n} \times \frac{J_r}{J_a} \times \frac{J_w}{SRF} \tag{4.2}$$

where

RQD is the Rock Quality Designation
J_n is the joint set number
J_r is the joint roughness number
J_a is the joint alteration number
J_w is the joint water reduction factor
SRF is the stress reduction factor

In explaining the meaning of the parameters used to determine the value of Q, Barton et al. (1974) offer the following comments:

The first quotient (RQD/J_n), representing the structure of the rock mass, is a crude measure of the block or particle size, with the two extreme values (100/0.5 and 10/20) differing by a factor of 400. If the quotient is interpreted in units of centimetres, the extreme 'particle sizes' of 200 to 0.5 cm are seen to be crude but fairly realistic approximations. Probably the largest blocks should be several times this size and the smallest fragments less than half the size. (Clay particles are of course excluded).

The second quotient (J_r/J_a) represents the roughness and frictional characteristics of the joint walls or filling materials. This quotient is weighted in favour of rough, unaltered joints in direct contact. It is to be expected that such surfaces will be close to peak strength, that they will dilate strongly when sheared, and they will therefore be especially favourable to tunnel stability.

When rock joints have thin clay mineral coatings and fillings, the strength is reduced significantly. Nevertheless, rock wall contact after small shear displacements have occurred may be a very important factor for preserving the excavation from ultimate failure.

Where no rock wall contact exists, the conditions are extremely unfavourable to tunnel stability. The 'friction angles' (given in Table 4.6) are a little below the residual strength values for most clays, and are possibly down-graded by the fact that these clay bands or fillings may tend to consolidate during shear, at least if normal consolidation or if softening and swelling has occurred. The swelling pressure of montmorillonite may also be a factor here.

The third quotient (J_w/SRF) consists of two stress parameters. SRF is a measure of: 1) loosening load in the case of an excavation through shear zones and clay bearing rock, 2) rock stress in competent rock, and 3) squeezing loads in plastic incompetent rocks. It can be regarded as a total stress parameter. The parameter J_w is a measure of water pressure, which has an adverse effect on the shear strength of joints due to a reduction in effective normal stress. Water may, in addition, cause softening and possible out-wash in the case of clay-filled joints. It has proved impossible to combine these two parameters in terms of inter-block effective stress, because paradoxically a high value of effective normal stress may sometimes signify less stable conditions than a low value, despite the higher shear strength. The quotient (J_w/SRF) is a complicated empirical factor describing the 'active stress'.

It appears that the rock tunnelling quality Q can now be considered to be a function of only three parameters which are crude measures of:

1. Block size (RQD/J_n)
2. Inter-block shear strength (J_r / J_a)
3. Active stress (J_w /SRF)

Undoubtedly, there are several other parameters which could be added to improve the accuracy of the classification system. One of these would be the joint orientation. Although many case records include the necessary information on structural orientation in relation to excavation axis, it was not found to be the important general parameter that might be expected. Part of the reason for this may be that the orientations of many types of excavations can be, and normally are, adjusted to avoid the maximum effect of unfavourably oriented major joints. However, this choice is not available in the case of tunnels, and more than half the case records were in this category. The parameters J_n, J_r and J_a appear to play a more important role than orientation, because the number of joint sets determines the degree of freedom for block movement (if any), and the frictional and dilational characteristics can vary more than the down-dip gravitational component of unfavourably oriented joints. If joint orientations had been included the classification would have been less general, and its essential simplicity lost.

Table 4.6 gives the classification of individual parameters used to obtain the Tunnelling Quality Index Q for a rock mass. The use of this table is illustrated in the following example.

A 15 m span crusher chamber for an underground mine is to be excavated in a norite at a depth of 2,100 m below surface. The rock mass contains two sets of joints controlling stability. These joints are undulating, rough and unweathered with very minor surface staining. *RQD* values range from 85% to 95% and laboratory tests on core samples of intact rock give an average uniaxial compressive strength of 170 MPa. The principal stress directions are approximately vertical and horizontal and the magnitude of the horizontal principal stress is approximately 1.5 times that of the vertical principal stress. The rock mass is locally damp but there is no evidence of flowing water.

The numerical value of *RQD* is used directly in the calculation of Q and, for this rock mass, an average value of 90 will be used. Table 4.6.2 shows that, for two joint sets, the joint set number, $J_n = 4$. For rough or irregular joints which are undulating, Table 4.6.3 gives a joint roughness number of $J_r = 3$. Table 4.6.4 gives the joint alteration number, $J_a = 1.0$, for unaltered joint walls with surface staining only. Table 4.6.5 shows that, for an excavation with minor inflow, the joint water reduction factor, $J_w = 1.0$. For a depth below surface of 2,100 m the overburden stress will be approximately 57 MPa and, in this case, the major principal stress $\sigma_1 = 85$ MPa. Since the uniaxial compressive strength of the norite is approximately 170 MPa, this gives a ratio of $\sigma_c / \sigma_1 = 2$. Table 4.6.6 shows that, for competent rock with rock stress problems, this value of σ_c/σ_1 can be expected to produce heavy rock burst conditions and that the value of *SRF* should lie between 10 and 20. A value of $SRF = 15$ will be assumed for this calculation. Using these values gives:

$$Q = \frac{90}{4} \times \frac{3}{1} \times \frac{1}{15} = 4.5$$

In relating the value of the index Q to the stability and support requirements of underground excavations, Barton et al. (1974) defined an additional parameter which they called the *Equivalent Dimension*, D_e, of the excavation. This dimension is obtained by dividing the span, diameter or wall height of the excavation by a quantity called the *Excavation Support Ratio*, *ESR*. Hence:

$$D_e = \frac{\text{Excavation span, diameter or height (m)}}{\text{Excavation Support Ratio } ESR}$$

The value of *ESR* is related to the intended use of the excavation and to the degree of security which is demanded of the support system installed to maintain the stability of the excavation. Barton et al. (1974) suggest the following values:

	Excavation category	*ESR*
A	Temporary mine openings	3-5
B	Permanent mine openings, water tunnels for hydro power (excluding high pressure penstocks), pilot tunnels, drifts and headings for large excavations	1.6
C	Storage rooms, water treatment plants, minor road and railway tunnels, surge chambers, access tunnels	1.3
D	Power stations, major road and railway tunnels, civil defence chambers, portal intersections	1.0
E	Underground nuclear power stations, railway stations, sports and public facilities, factories	0.8

The crusher station discussed above falls into the category of permanent mine openings and is assigned an excavation support ratio *ESR* = 1.6. Hence, for an excavation span of 15 m, the equivalent dimension, $D_e = 15/1.6 = 9.4$.

The equivalent dimension, *De*, plotted against the value of *Q*, is used to define a number of support categories in a chart published in the original paper by Barton et al. (1974). This chart has recently been updated by Grimstad and Barton (1993) to reflect the increasing use of steel fibre reinforced shotcrete in underground excavation support. Figure 4.3 is reproduced from this updated chart.

From Figure 4.3, a value of D_e of 9.4 and a value of *Q* of 4.5 places this crusher excavation in category (4) which requires a pattern of rockbolts (spaced at 2.3 m) and 40 to 50 mm of unreinforced shotcrete.

Because of the mild to heavy rock burst conditions which are anticipated, it may be prudent to destress the rock in the walls of this crusher chamber. This is achieved by using relatively heavy production blasting to excavate the chamber and omitting the smooth blasting usually used to trim the final walls of an excavation such as an underground powerhouse at shallower depth. Caution is recommended in the use of destress blasting and, for critical applications, it may be advisable to seek the advice of a blasting specialist before embarking on this course of action.

Løset (1992) suggests that, for rocks with $4 < Q < 30$, blasting damage will result in the creation of new 'joints' with a consequent local reduction in the value of *Q* for the rock surrounding the excavation. He suggests that this can be accounted for by reducing the *RQD* value for the blast damaged zone.

Assuming that the *RQD* value for the destressed rock around the crusher chamber drops to 50%, the resulting value of $Q = 2.9$. From Figure 4.3, this value of *Q*, for an equivalent dimension, D_e of 9.4, places the excavation just inside category (5) which requires rockbolts, at approximately 2 m spacing, and a 50 mm thick layer of steel fibre reinforced shotcrete.

Barton et al. (1980) provide additional information on rockbolt length, maximum unsupported spans and roof support pressures to supplement the support recommendations published in the original 1974 paper.

The length *L* of rockbolts can be estimated from the excavation width *B* and the Excavation Support Ratio *ESR*:

$$L = \frac{2 + 0.15B}{ESR} \tag{4.3}$$

Table 4.6: Classification of individual parameters used in the Tunnelling Quality Index Q (After Barton et al., 1974).

DESCRIPTION	VALUE	NOTES
1. ROCK QUALITY DESIGNATION	*RQD*	
A. Very poor	0-25	1. Where *RQD* is reported or measured as $\leq$ 10 (including 0),
B. Poor	25-50	a nominal value of 10 is used to evaluate Q.
C. Fair	50-75	
D. Good	75-90	2. RQD intervals of 5, i.e. 100, 95, 90 etc. are sufficiently
E. Excellent	90-100	accurate.
2. JOINT SET NUMBER	J_n	
A. Massive, no or few joints	0.5-1.0	
B. One joint set	2	
C. One joint set plus random	3	
D. Two joint sets	4	
E. Two joint sets plus random	6	
F. Three joint sets	9	1. For intersections use $(3.0 \times J_n)$.
G. Three joint sets plus random	12	
H. Four or more joint sets, random, heavily jointed, 'sugar cube', etc.	15	2. For portals use $(2.0 \times J_n)$.
J. Crushed rock, earthlike	20	
3. JOINT ROUGHNESS NUMBER	J_r	
a. Rock wall contact		
b. Rock wall contact before 10 cm shear		
A. Discontinuous joints	4	
B. Rough and irregular, undulating	3	
C. Smooth undulating	2	
D. Slickensided undulating	1.5	1. Add 1.0 if the mean spacing of the relevant joint set is
E. Rough or irregular, planar	1.5	greater than 3 m.
F. Smooth, planar	1.0	
G. Slickensided, planar	0.5	2. J_r = 0.5 can be used for planar, slickensided joints having
c. No rock wall contact when sheared		lineations, provided that the lineations are oriented for
H. Zones containing clay minerals thick enough to prevent rock wall contact	1.0 (nominal)	minimum strength.
J. Sandy, gravely or crushed zone thick enough to prevent rock wall contact	1.0 (nominal)	

DESCRIPTION	VALUE	ϕr degrees (approx.)	NOTES
4. JOINT ALTERATION NUMBER	J_a	ϕr degrees (approx.)	
a. Rock wall contact			
A. Tightly healed, hard, non-softening, impermeable filling	0.75		1. Values of ϕr, the residual friction angle, are intended as an approximate guide
B. Unaltered joint walls, surface staining only	1.0	25-35	to the mineralogical properties of the
C. Slightly altered joint walls, non-softening mineral coatings, sandy particles, clay-free disintegrated rock, etc.	2.0	25-30	alteration products, if present.
D. Silty-, or sandy-clay coatings, small clay-fraction (non-softening)	3.0	20-25	
E. Softening or low-friction clay mineral coatings, i.e. kaolinite, mica. Also chlorite, talc, gypsum and graphite etc., and small quantities of swelling clays. (Discontinuous coatings, 1-2 mm or less in thickness)	4.0	8-16	

Table 4.6: (continued).

DESCRIPTION	VALUE	NOTES	
4, JOINT ALTERATION NUMBER	J_a	ϕr degrees (approx.)	
b. Rock wall contact before 10 cm shear			
F. Sandy particles, clay-free, disintegrating rock etc.	4.0	25-30	
G. Strongly over-consolidated, non-softening clay mineral fillings (continuous < 5 mm thick)	6.0	16-24	
H. Medium or low over-consolidation, softening clay mineral fillings (continuous < 5 mm thick)	8.0	12-16	
J. Swelling clay fillings, i.e. montmorillonite, (continuous < 5 mm thick). Values of J_a depend on percent of swelling clay-size particles, and access to water.	8.0-12.0	6-12	
c. No rock wall contact when sheared			
K. Zones or bands of disintegrated or crushed	6.0		
L. rock and clay (see G, H and J for clay	8.0		
M. conditions)	8.0-12.0	6-24	
N. Zones or bands of silty- or sandy-clay, small clay fraction, non-softening	5.0		
O. Thick continuous zones or bands of clay	10.0-13.0		
P. & R. (see G.H and J for clay conditions)	6.0-24.0		
5. JOINT WATER REDUCTION	J_w	approx. water pressure (kgf/cm^2)	
A. Dry excavation or minor inflow i.e. < 5 l/m locally	1.0	< 1.0	
B. Medium inflow or pressure, occasional outwash of joint fillings	0.66	1.0-2.5	
C. Large inflow or high pressure in competent rock with unfilled joints	0.5	2.5-10.0	1. Factors C to F are crude estimates; increase J_w if drainage installed.
D. Large inflow or high pressure	0.33	2.5-10.0	
E. Exceptionally high inflow or pressure at blasting, decaying with time	0.2-0.1	> 10	2. Special problems caused by ice formation are not considered.
F. Exceptionally high inflow or pressure	0.1-0.05	> 10	
6. STRESS REDUCTION FACTOR		*SRF*	
a. Weakness zones intersecting excavation, which may cause loosening of rock mass when tunnel is excavated			
A. Multiple occurrences of weakness zones containing clay or chemically disintegrated rock, very loose surrounding rock any depth)		10.0	1. Reduce these values of *SRF* by 25-50% if the relevant shear zones only influence but do not intersect the excavation.
B. Single weakness zones containing clay, or chemically disintegrated rock (excavation depth < 50 m)		5.0	
C. Single weakness zones containing clay, or chemically disintegrated rock (excavation depth > 50 m)		2.5	
D. Multiple shear zones in competent rock (clay free), loose surrounding rock (any depth)		7.5	
E. Single shear zone in competent rock (clay free). (depth of excavation < 50 m)		5.0	
F. Single shear zone in competent rock (clay free). (depth of excavation > 50 m)		2.5	
G. Loose open joints, heavily jointed or 'sugar cube', (any depth)		5.0	

Table 4.6: (continued) .

DESCRIPTION			VALUE	NOTES
6. STRESS REDUCTION FACTOR			*SRF*	
b. Competent rock, rock stress problems				
	σ_c/σ_1	$\sigma_t\sigma_1$		2. For strongly anisotropic virgin stress field
H. Low stress, near surface	> 200	> 13	2.5	(if measured): when $5 \leq \sigma_1/\sigma_3 \leq 10$, reduce σ_c
J. Medium stress	200-10	13-0.66	1.0	to $0.8\sigma_c$ and σ_t to $0.8\sigma_t$. When $\sigma_1/\sigma_3 > 10$,
K. High stress, very tight structure	10-5	0.66-0.33	0.5-2	reduce σ_c and σ_t to $0.6\sigma_c$ and $0.6\sigma_t$, where
(usually favourable to stability, may				σ_c = unconfined compressive strength, and
be unfavourable to wall stability)				σ_t = tensile strength (point load) and σ_1 and
L. Mild rockburst (massive rock)	5-2.5	0.33-0.16	5-10	σ_3 are the major and minor principal stresses.
M. Heavy rockburst (massive rock)	< 2.5	< 0.16	10-20	3. Few case records available where depth of
c. Squeezing rock, plastic flow of incompetent rock				crown below surface is less than span width.
under influence of high rock pressure				Suggest *SRF* increase from 2.5 to 5 for such
N. Mild squeezing rock pressure			5-10	cases (see H).
O. Heavy squeezing rock pressure			10-20	
d. Swelling rock, chemical swelling activity depending on presence of water				
P. Mild swelling rock pressure			5-10	
R. Heavy swelling rock pressure			10-15	

ADDITIONAL NOTES ON THE USE OF THESE TABLES

When making estimates of the rock mass Quality (Q), the following guidelines should be followed in addition to the notes listed in the tables:

1. When borehole core is unavailable, RQD can be estimated from the number of joints per unit volume, in which the number of joints per metre for each joint set are added. A simple relationship can be used to convert this number to RQD for the case of clay free rock masses: $RQD = 115\text{-}3.3\,J_v$ (approx.), where J_v = total number of joints per m^3 ($0 < RQD < 100$ for $35 > J_v > 4.5$).

2. The parameter J_n representing the number of joint sets will often be affected by foliation, schistosity, slaty cleavage or bedding etc. If strongly developed, these parallel 'joints' should obviously be counted as a complete joint set. However, if there are few 'joints' visible, or if only occasional breaks in the core are due to these features, then it will be more appropriate to count them as 'random' joints when evaluating J_n.

3. The parameters J_r and J_a (representing shear strength) should be relevant to the weakest significant joint set or clay filled discontinuity in the given zone. However, if the joint set or discontinuity with the minimum value of J_r/J_a is favourably oriented for stability, then a second, less favourably oriented joint set or discontinuity may sometimes be more significant, and its higher value of J_r/J_a should be used when evaluating Q. The value of J_r/J_a should in fact relate to the surface most likely to allow failure to initiate.

4. When a rock mass contains clay, the factor *SRF* appropriate to loosening loads should be evaluated. In such cases the strength of the intact rock is of little interest. However, when jointing is minimal and clay is completely absent, the strength of the intact rock may become the weakest link, and the stability will then depend on the ratio rock-stress/rock-strength. A strongly anisotropic stress field is unfavourable for stability and is roughly accounted for as in note 2 in the table for stress reduction factor evaluation.

5. The compressive and tensile strengths (σ_c and σ_t) of the intact rock should be evaluated in the saturated condition if this is appropriate to the present and future in situ conditions. A very conservative estimate of the strength should be made for those rocks that deteriorate when exposed to moist or saturated conditions.

The maximum unsupported span can be estimated from:

$$\text{Maximum span (unsupported)} = 2\,ESR\,Q^{0.4} \qquad (4.4)$$

Based upon analyses of case records, Grimstad and Barton (1993) suggest that the relationship between the value of Q and the permanent roof support pressure P_{roof} is estimated from:

$$P_{roof} = \frac{2\sqrt{J_n}\,Q^{-\frac{1}{3}}}{3J_r} \qquad (4.5)$$

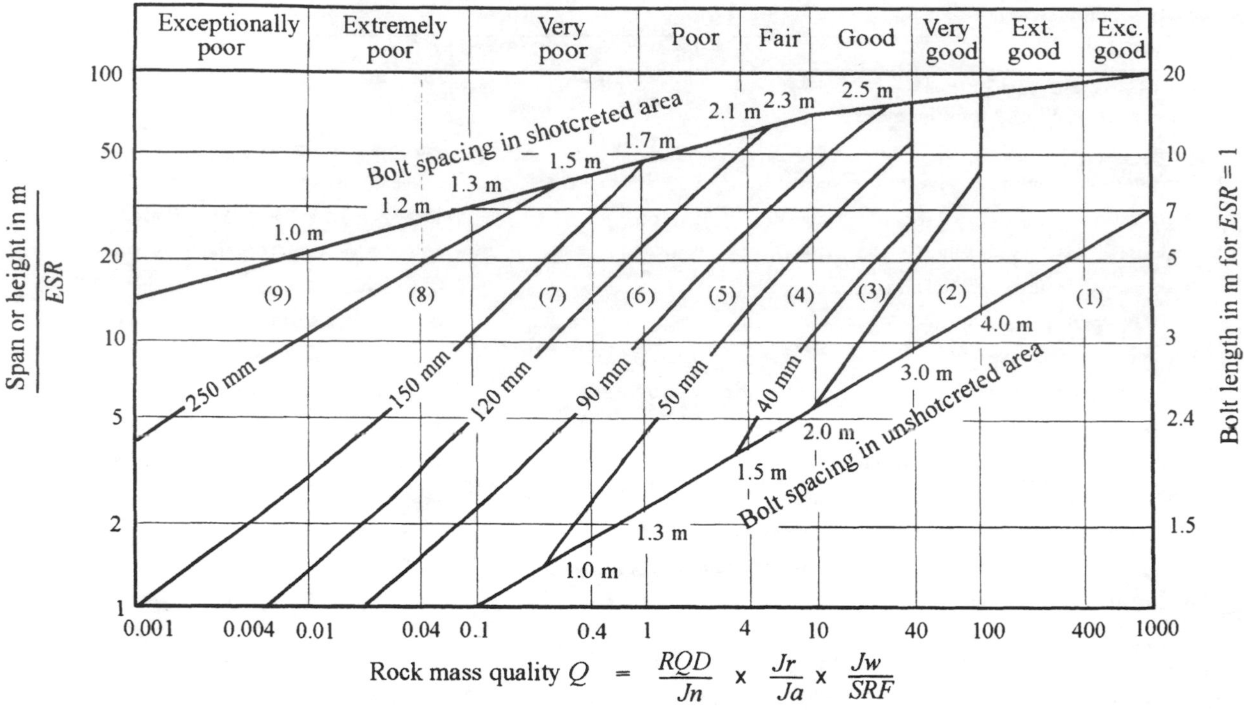

REINFORCEMENT CATEGORIES

1) Unsupported
2) Spot bolting
3) Systematic bolting
4) Systematic bolting with 40-100 mm unreinforced shotcrete
5) Fibre reinforced shotcrete, 50 - 90 mm, and bolting
6) Fibre reinforced shotcrete, 90 - 120 mm, and bolting
7) Fibre reinforced shotcrete, 120 - 150 mm, and bolting
8) Fibre reinforced shotcrete, > 150 mm, with reinforced ribs of shotcrete and bolting
9) Cast concrete lining

Figure 4.3: Estimated support categories based on the tunnelling quality index Q (After Grimstad and Barton, 1993).

4.6 Using rock mass classification systems

The two most widely used rock mass classifications are Bieniawski's *RMR* (1976, 1989) and Barton et al.'s *Q* (1974). Both methods incorporate geological, geometric and design/engineering parameters in arriving at a quantitative value of their rock mass quality. The similarities between *RMR* and *Q* stem from the use of identical, or very similar, parameters in calculating the final rock mass quality rating. The differences between the systems lie in the different weightings given to similar parameters and in the use of distinct parameters in one or the other scheme.

RMR uses compressive strength directly while *Q* only considers strength as it relates to in situ stress in competent rock. Both schemes deal with the geology and geometry of the rock mass, but in slightly different ways. Both consider groundwater, and both include some component of rock material strength. Some estimate of orientation can be incorporated into *Q* using a guideline presented by Barton et al. (1974): 'the parameters J_r and J_a should .. relate to the surface most likely to allow failure to initiate.' The greatest difference between the two systems is the lack of a stress parameter in the *RMR* system.

When using either of these methods, two approaches can be taken. One is to evaluate the rock mass specifically for the parameters included in the classification methods; the other is to accurately characterise the rock mass and then attribute parameter ratings at a later time. The latter method is recommended since it gives a full and complete description of the rock mass which can easily be translated into either classification index. If rating values alone had been recorded during mapping, it would be almost impossible to carry out verification studies.

In many cases, it is appropriate to give a range of values to each parameter in a rock mass classification and to evaluate the significance of the final result. An example of this approach is given in Figure 4.4 which is reproduced from field notes prepared by Dr. N. Barton on a project. In this particular case, the rock mass is dry and is subjected to 'medium' stress conditions (Table 4.6.6.K) and hence $J_w = 1.0$ and $SRF = 1.0$. Histograms showing the variations in *RQD*, J_n, J_r and J_a,along the exploration adit mapped, are presented in this figure. The average value of $Q = 8.9$ and the approximate range of Q is $1.7 < Q < 20$. The average value of Q can be used in choosing a basic support system while the range gives an indication of the possible adjustments which will be required to meet different conditions encountered during construction.

A further example of this approach is given in a paper by Barton et al. (1992) concerned with the design of a 62 m span underground sports hall in jointed gneiss. Histograms of all the input parameters for the Q system are presented and analysed in order to determine the weighted average value of Q.

Carter (1992) has adopted a similar approach, but extended his analysis to include the derivation of a probability distribution function and the calculation of a probability of failure in a discussion on the stability of surface crown pillars in abandoned metal mines.

Throughout this chapter it has been suggested that the user of a rock mass classification scheme should check that the latest version is being used. An exception is the use of Bieniawski's *RMR* classification for rock mass strength estimates (discussed in Chapter 8) where the 1976 version as well as the 1989 version are used. It is also worth repeating that the use of two rock mass classification schemes is advisable.

4.7 Estimation of in situ deformation modulus

The in situ deformation modulus of a rock mass is an important parameter in any form of numerical analysis and in the interpretation of monitored deformation around underground openings. Since this parameter is very difficult and expensive to determine in the field, several attempts have been made to develop methods for estimating its value, based upon rock mass classifications.

In the 1960s several attempts were made to use Deere's *RQD* for estimating in situ deformation modulus, but this approach is seldom used today (Deere and Deere, 1988).

Bieniawski (1978) analysed a number of case histories and proposed the following relationship for estimating the in situ deformation modulus, E_m, from *RMR*:

$$E_m = 2\,RMR - 100 \tag{4.6}$$

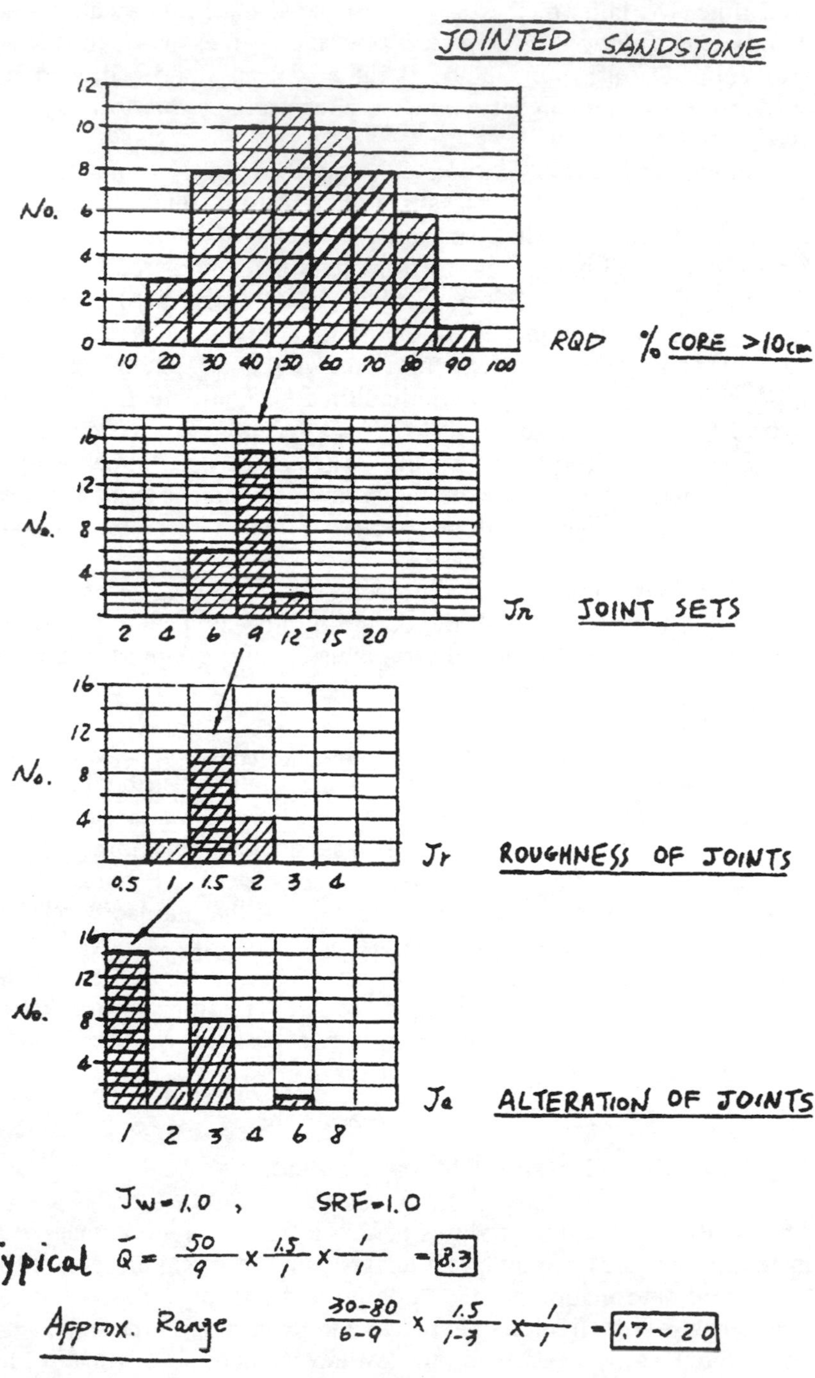

Figure 4.4: Histograms showing variations in *RQD*, J_n, J_r and J_a for a dry jointed sandstone under 'medium' stress conditions, reproduced from field notes prepared by Dr. N. Barton.

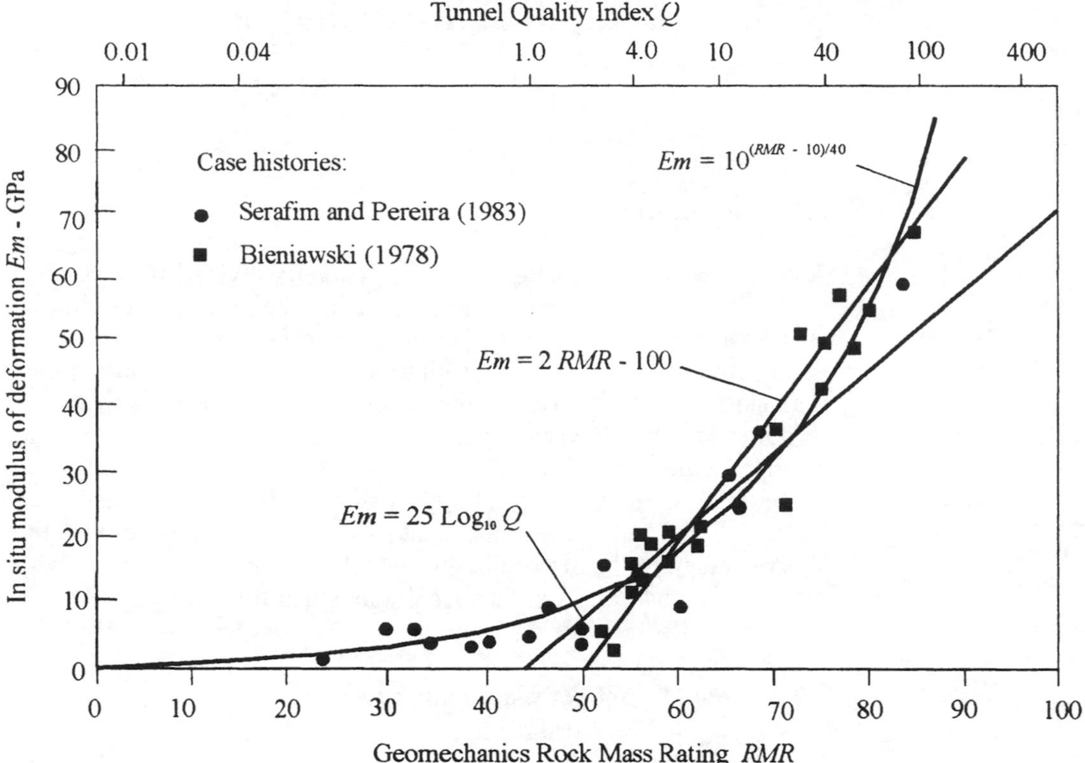

Figure 4.5: Prediction of in situ deformation modulus *Em* from rock mass claffications.

Based on the analyses of a number of case histories, many of which involved dam foundations for which the deformation modulii were evaluated by back analysis of measured deformations, Serafim and Pereira (1983) proposed the following relationship between E_m and *RMR*:

$$E_m = 10^{\frac{(RMR-10)}{40}} \tag{4.7}$$

More recently Barton et al. (1980), Barton et al. (1992) and Grimstad and Barton (1993) have found good agreement between measured displacements and predictions from numerical analyses using in situ deformation modulus values estimated from:

$$E_m = 25 \, \text{Log}_{10} Q \tag{4.8}$$

Curves defined by equations 4.6, 4.7 and 4.8, together with the case history observations of Bieniawski (1978) and Serafim and Pereira (1983) are plotted in Figure 4.5. This figure suggests that equation 4.7 provides a reasonable fit for all of the observations plotted and it has the advantage of covering a wider range of *RMR* values than either of the other two equations.

5 Shear strength of discontinuities

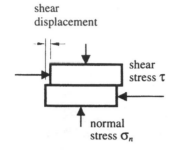

5.1 Introduction

A hard rock mass at shallow depth is generally divided into discrete blocks by intersecting discontinuities such as bedding planes, joints, shear zones and faults. Since the in situ stresses are low at shallow depth, stress induced failure of intact rock material is usually minimal and plays a minor role in the behaviour of the rock mass, which is dominated by gravity driven sliding on the discontinuities and rotation of the individual rock blocks.

In order to analyse the stability of this system of individual rock blocks, it is necessary to understand the factors which control the shear strength of the discontinuities which separate the blocks. These questions are addressed in the discussion which follows.

shear displacement

shear stress τ

normal stress σ_n

5.2 Shear strength of planar surfaces

Suppose that a number of samples of a rock, such as slate, are obtained for shear testing. Each sample contains a through-going bedding plane which is cemented; in other words, a tensile force would have to be applied to the two halves of the specimen in order to separate them. The bedding plane is absolutely planar, having no surface irregularities or undulations. As illustrated in the margin sketch, in a shear test each specimen is subjected to a stress σ_n normal to the bedding plane, and the shear stress τ, required to cause a displacement δ, is measured.

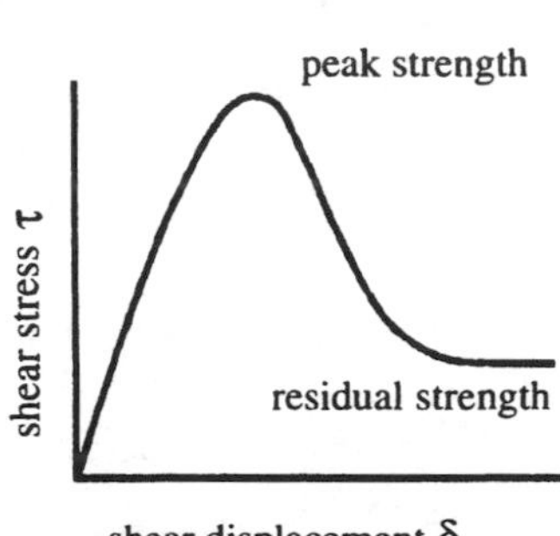

The shear stress will increase steeply until the peak strength is reached. This corresponds to the failure of the cementing material bonding the two halves of the bedding plane together. As the displacement continues, the shear stress will drop to some residual value which will then remain constant, even for large shear displacements.

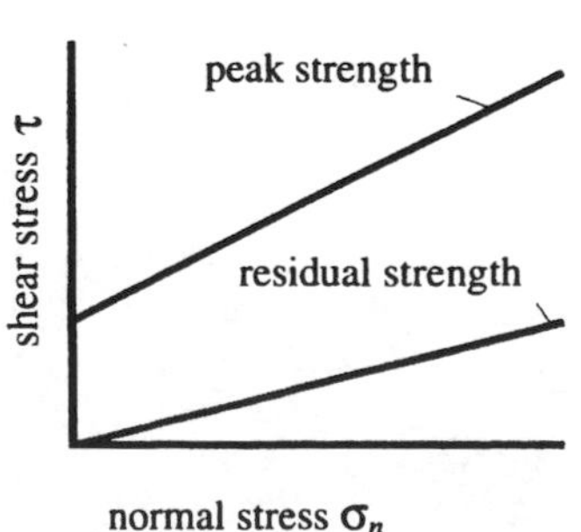

Plotting the peak and residual shear strengths for different normal stresses results in the envelopes illustrated in the lower margin sketch. For planar discontinuity surfaces, such as those considered in this example, the experimental points will generally fall along straight lines. The relationship between the peak shear strength τ_p and the normal stress σ_n can be represented by the Mohr-Coulomb equation:

$$\tau_p = c + \sigma_n \tan\phi \qquad (5.1)$$

where c is the cohesive strength of the cemented surface and
ϕ is the angle of friction.

In the case of the residual strength, the cohesion c has dropped to zero and the relationship between τ_r and σ_n can be represented by:

$$\tau_r = \sigma_n \tan\phi_r \qquad (5.2)$$

where ϕ_r is the residual angle of friction.

This example has been discussed in order to illustrate the physical meaning of the term *cohesion*, a soil mechanics term, which has been adopted by the rock mechanics community. In shear tests on soils, the stress levels are generally an order of magnitude lower than those involved in rock testing and the cohesive strength of a soil is a result of the adhesion of the soil particles. In rock mechanics, true cohesion occurs when cemented surfaces are sheared. However, in many practical applications, the term cohesion is used for convenience and it refers to a mathematical quantity related to surface roughness, as discussed below. Cohesion is simply the intercept on the τ axis at zero normal stress, i.e., the cohesion intercept.

A quantity, which is fundamental to the understanding of the shear strength of discontinuity surfaces, is the *basic* friction angle ϕ_b. This is approximately equal to the residual friction angle ϕ_r but it is generally measured by testing sawn or ground rock surfaces. These tests, which can be carried out on surfaces as small as 50 mm × 50 mm, will produce a straight line plot defined by the equation:

$$\tau = \sigma_n \tan\phi_b \qquad (5.3)$$

5.3 Shear strength of rough surfaces

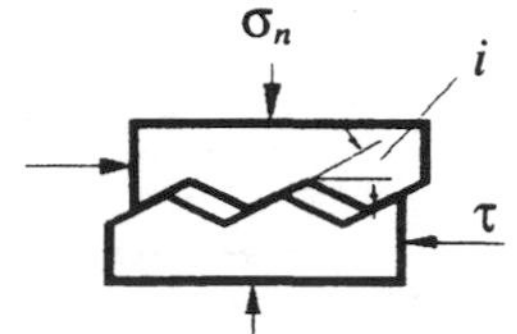

A natural discontinuity surface in hard rock is never as smooth as a sawn or ground surface of the type used for determining the basic friction angle. The undulations and asperities on a natural joint surface have a significant influence on its shear behaviour. Generally, this surface roughness increases the shear strength of the surface, and this strength increase is extremely important in terms of the stability of underground openings.

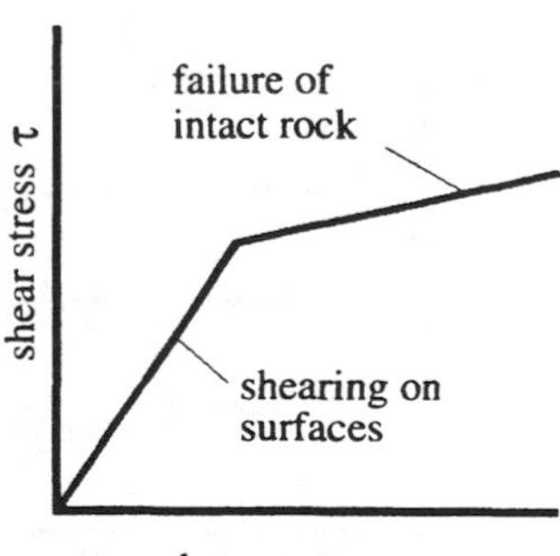

Patton (1966) demonstrated this influence by means of a simple experiment in which he carried out shear tests on 'saw-tooth' specimens such as the one illustrated in the margin sketch. Shear displacement in these specimens occurs as a result of the surfaces moving up the inclined faces, causing *dilation* (an increase in volume) of the specimen.

The shear strength of Patton's saw-tooth specimens can be represented by the equation:

$$\tau = \sigma_n \tan(\phi_b + i) \qquad (5.4)$$

where ϕ_b is the basic friction angle of the surface and
i is the angle of the saw-tooth face.

This equation is valid at low normal stresses where shear displacement is due to sliding along the inclined surfaces. At higher normal stresses, the strength of the intact material will be exceeded and the teeth will tend to break off, resulting in a shear strength behaviour which is more closely related to the intact material strength than to the frictional characteristics of the surfaces.

Barton and his co-workers (1973, 1976, 1977, 1990) have studied the behaviour of natural rock joints in great detail and have proposed that Equation 5.4 can be re-written as:

$$\tau = \sigma_n \tan\left[\phi_b + JRC \log_{10}\left(\frac{JCS}{\sigma_n}\right)\right] \qquad (5.5)$$

where *JRC* is the joint roughness coefficient and
JCS is the joint wall compressive strength.

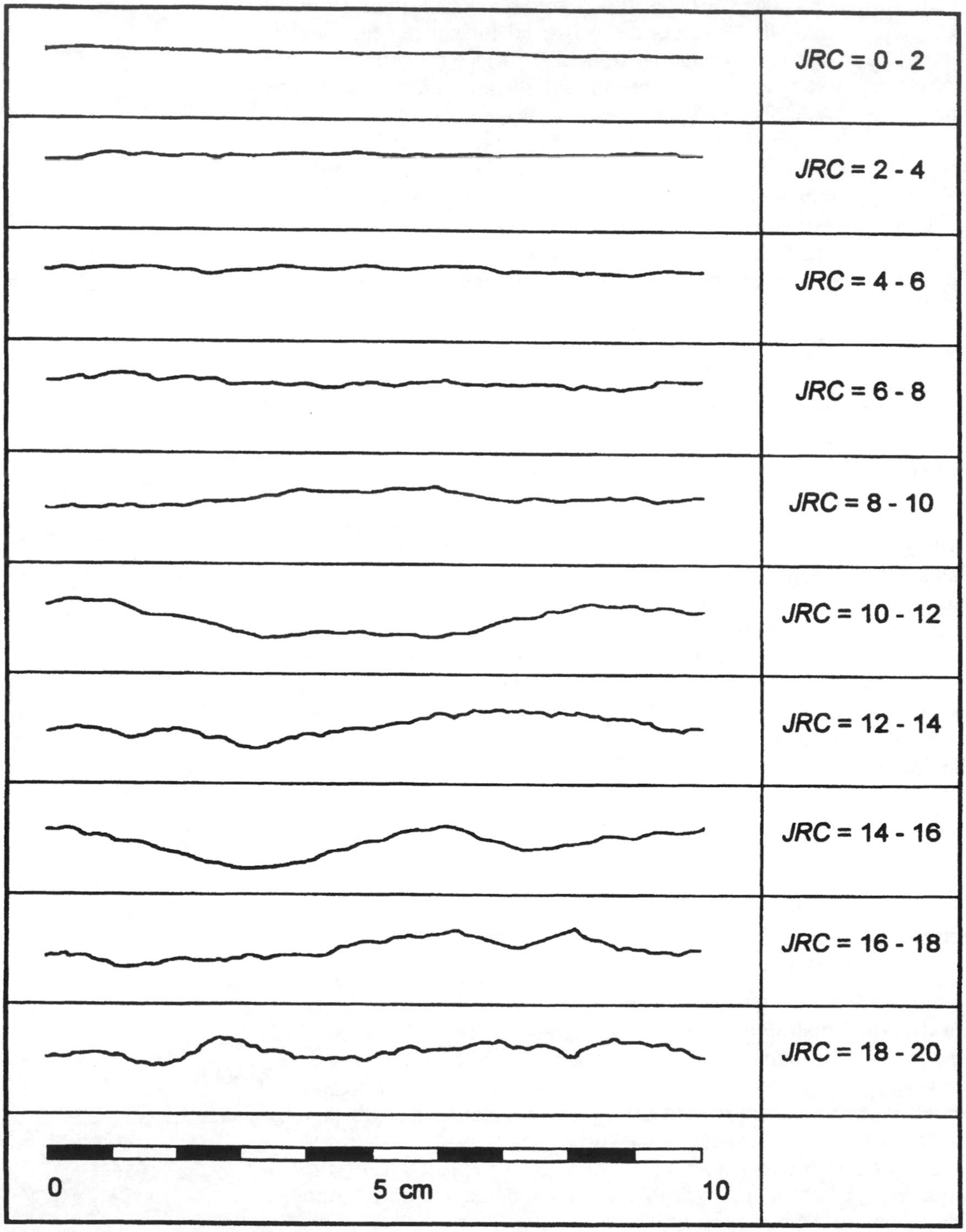

JRC = 0 - 2

JRC = 2 - 4

JRC = 4 - 6

JRC = 6 - 8

JRC = 8 - 10

JRC = 10 - 12

JRC = 12 - 14

JRC = 14 - 16

JRC = 16 - 18

JRC = 18 - 20

0 5 cm 10

Figure 5.1: Roughness profiles and corresponding *JRC* values (After Barton and Choubey, 1977).

Description	Profile	J_r	*JRC* 200mm	*JRC* 1 m
Rough		4	20	11
Smooth		3	14	9
Slickensided	Stepped	2	11	8
Rough		3	14	9
Smooth		2	11	8
Slickensided	Undulating	1.5	7	6
Rough		1.5	2.5	2.3
Smooth		1.0	1.5	0.9
Slickensided	Planar	0.5	0.5	0.4

Figure 5.2: Relationship between J_r in the Q system and *JRC* for 200 mm and 1000 mm samples (After Barton, 1987).

5.3.1 *Field estimates of JRC*

The joint roughness coefficient *JRC* is a number which is determined by comparing the appearance of a discontinuity surface with standard profiles published by Barton and others. One of the most useful of these profile sets was published by Barton and Choubey (1977) and is reproduced in Figure 5.1. Note that these profiles have been reproduced at full scale in order to facilitate direct comparison with measured roughness profiles, where these are available.

Barton (1987) published a table relating J_r to *JRC* and this table is reproduced in Figure 5.2.

Barton and Bandis (1990) suggest that *JRC* can also be estimated from a simple tilt test in which a pair of matching discontinuity surfaces are tilted until one slides on the other. The *JRC* value is estimated from the tilt angle α by means of the following equation.

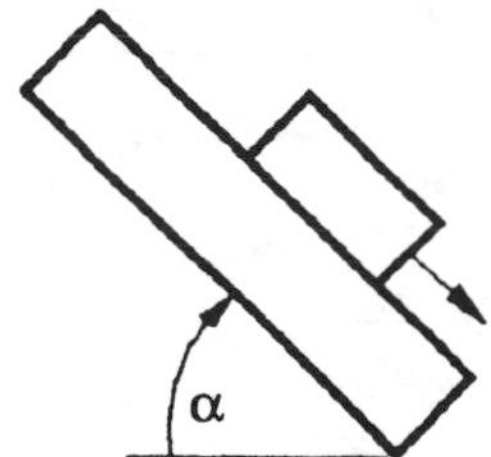

$$JRC = \frac{\alpha - \phi_b}{\log_{10}\left[\frac{JCS}{\sigma_n}\right]} \quad (5.6)$$

For small samples, the normal stress σ_n may be as low as 0.001 MPa. Assuming this value for a typical case in which the tilt angle $\alpha = 65°$, the basic friction angle $\phi_b = 30°$ and the joint wall compressive strength $JCS = 100$ MPa, Equation 5.6 gives $JRC = 7$.

5.3.2 *Field estimates of JCS*

Suggested methods for estimating the joint wall compressive strength were published by the ISRM (1978). The use of the Schmidt rebound hammer for estimating joint wall compressive strength was proposed by Deere and Miller (1966).

5.3.3 *Influence of scale on JRC and JCS*

Equation 5.5 suggests that there are three factors which control the shear strength of natural discontinuities: the basic friction angle ϕ_b, a geometrical component *JRC*, and an asperity failure component controlled by the ratio (JCS/σ_n). Figure 5.3, adapted from a figure originally published by Bandis (1980), shows that, as the scale increases, the effective roughness of the surface (*JRC*) decreases. Hence the shear strength of the surface decreases. Also, because of the greater possibility of weaknesses in a large surface, it is also likely that the average joint wall compressive strength (*JCS*) decreases with increasing scale.

On the basis of extensive testing of joints, joint replicas, and a review of literature, Barton and Bandis (1982) proposed the scale corrections for *JRC* and *JCS* defined by Equations 5.7 and 5.8.

$$JRC_n = JRC_o\left[\frac{L_n}{L_o}\right]^{-0.02\,JRC_o} \quad (5.7)$$

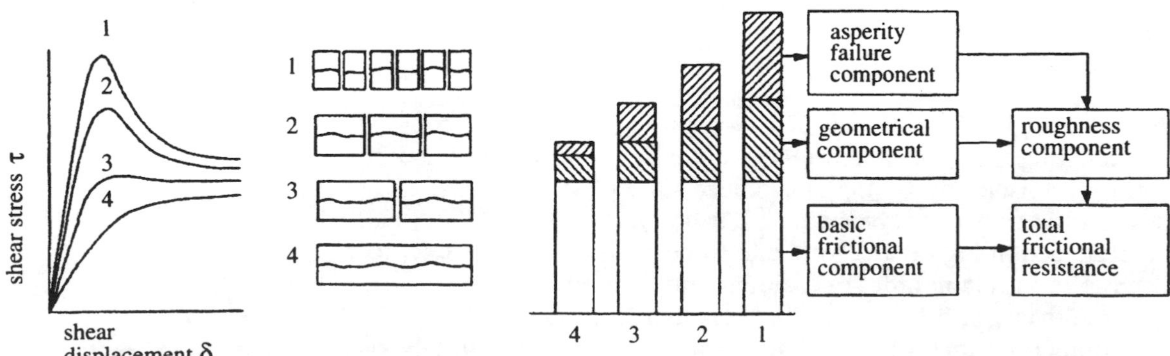

Figure 5.3: Influence of scale on the three components of the shear strength of a rough discontinuity. After Bandis (1990) and Barton and Bandis (1990).

$$JCS_n = JCS_o \left[\frac{L_n}{L_o} \right]^{-0.03\, JCS_o} \tag{5.8}$$

where JRC_o, JCS_o and L_o (length) refer to 100 mm laboratory scale samples and JRC_n, JCS_n and L_n refer to in situ block sizes.

The quantity JCS_o, the joint wall compressive strength of a 100 mm laboratory specimen, has a maximum value equal to the uniaxial compressive strength of the intact rock material. This maximum value will be found in the case of fresh, unweathered or unaltered discontinuity surfaces. The strength will be reduced by weathering or alteration of the surface and also by the size of the surface, as suggested by Equation 5.8.

5.4 Shear strength of filled discontinuities

The discussion presented in the previous sections has dealt with the shear strength of discontinuities in which rock wall contact occurs over the entire length of the surface under consideration. This shear strength can be reduced drastically when part or all of the surface is not in intimate contact, but covered by soft filling material such as clay gouge. For planar surfaces, such as bedding planes in sedimentary rock, a thin clay coating will result in a significant shear strength reduction. For a rough or undulating joint, the filling thickness has to be greater than the amplitude of the undulations before the shear strength is reduced to that of the filling material.

A comprehensive review of the shear strength of filled discontinuities was prepared by Barton (1974) and a summary of the shear strengths of typical discontinuity fillings, based on Barton s review, is given in Table 5.1.

Where a significant thickness of clay or gouge fillings occurs in rock masses and where the shear strength of the filled discontinuities is likely to play an important role in the stability of the rock mass, it is strongly recommended that samples of the filling be sent to a soil mechanics laboratory for testing..

5.5 Influence of water pressure

When water pressure is present in a rock mass, the surfaces of the discontinuities are forced apart and the normal stress σn is reduced. Under steady state conditions, where there is sufficient time for the water pressures in the rock mass to reach equilibrium, the reduced normal stress is defined by $\sigma_n' = (\sigma_n - u)$, where u is the water pressure. The reduced normal stress σ_n' is usually called the *effective normal stress*, and it can be used in place of the normal stress term σ_n in all of the equations presented in previous sections of this chapter.

Table 5.1: Shear strength of filled discontinuities and filling materials (After Barton, 1974).

Rock	Description	Peak c' (MPa)	Peak $\phi°$	Residual c' (MPa)	Residual $\phi°$
Basalt	Clayey basaltic breccia, wide variation from clay to basalt content	0.24	42		
Bentonite	Bentonite seam in chalk	0.015	7.5		
	Thin layers	0.09-0.12	12-17		
	Triaxial tests	0.06-0.1	9-13		
Bentonic shale	Triaxial tests	0-0.27	8.5-29		
	Direct shear tests			0.03	8.5
Clays	Over-consolidated, slips, joints and minor shears	0-0.18	12-18.5	0-0.003	10.5-16
Clay shale	Triaxial tests	0.06	32		
	Stratification surfaces			0	19-25
Coal measure rocks	Clay mylonite seams, 10 to 25 mm	0.012	16	0	11-11.5
Dolomite	Altered shale bed, ± 150 mm thick	0.04	14.5	0.02	17
Diorite, granodiorite and porphyry	Clay gouge (2% clay, PI = 17%)	0	26.5		
Granite	Clay filled faults	0-0.1	24-45		
	Sandy loam fault filling	0.05	40		
	Tectonic shear zone, schistose and broken granites, disintegrated rock and gouge	0.24	42		
Greywacke	1-2 mm clay in bedding planes			0	21
Limestone	6 mm clay layer			0	13
	10-20 mm clay fillings	0.1	13-14		
	<1 mm clay filling	0.05-0.2	17-21		
Limestone, marl and lignites	Interbedded lignite layers	0.08	38		
	Lignite/marl contact	0.1	10		
Limestone	Marlaceous joints, 20 mm thick	0	25	0	15-24
Lignite	Layer between lignite and clay	0.014-.03	15-17.5		
Montmorillonite Bentonite clay	80 mm seams of bentonite (montmorillonite) clay in chalk	0.36 0.016-.02	14 7.5-11.5	0.08	11
Schists, quartzites and siliceous schists	100-15- mm thick clay filling	0.03-0.08	32		
	Stratification with thin clay	0.61-0.74	41		
	Stratification with thick clay	0.38	31		
Slates	Finely laminated and altered	0.05	33		
Quartz / kaolin / pyrolusite	Remoulded triaxial tests	0.042-.09	36-38		

5.6 Instantaneous cohesion and friction

Due to the historical development of the subject of rock mechanics, many of the analyses, used to calculate factors of safety against sliding, are expressed in terms of the Mohr-Coulomb cohesion (c) and friction angle (ϕ), defined in Equation 5.1. Since the 1970s it has been recognised that the relationship between shear strength and normal stress is more accurately represented by a non-linear relationship such as that proposed by Barton (1973). However, because such a

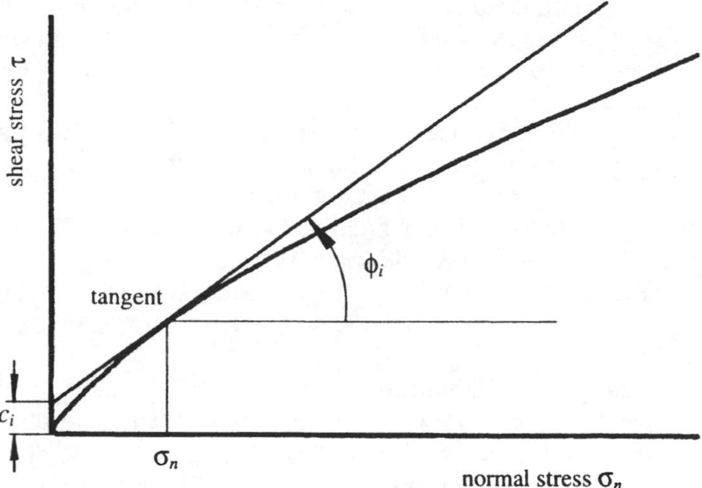

Figure 5.4: Definition of instantaneous cohesion c_i and instantaneous friction angle ϕ_i for a non-linear failure criterion.

Barton shear failure criterion

Input parameters:

Basic friction angle (PHIB)-degrees	29
Joint roughness coefficient (JRC)	16.9
Joint compressive strength (JCS)	96
Minimum normal stress (SIGNMIN)	0.360

Normal stress (SIGN) MPa	Shear strength (TAU) MPa	dTAU / dSIGN (DTDS)	Friction angle (PHI) degrees	Cohesive strength (COH) MPa
0.360	0.989	1.652	58.82	0.394
0.720	1.538	1.423	54.91	0.513
1.440	2.476	1.213	50.49	0.730
2.880	4.073	1.030	45.85	1.107
5.759	6.779	0.872	41.07	1.760
11.518	11.344	0.733	36.22	2.907
23.036	18.973	0.609	31.33	4.953
46.073	31.533	0.496	26.40	8.666

Cell formulae:

```
SIGNMIN=10^(LOG(JCS)-(70-PHIB)/JRC)
TAU = SIGN*TAN((PHIB+JRC*LOG(JCS/SIGN))*PI()/180)

DTDS =TAN((JRC*LOG(JCS/SIGN)+PHIB)*PI()/180)-
    (JRC/LN(10))*(TAN((JRC*LOG(JCS/SIGN)+PHIB)*PI()/180)^2+1)*
PI()/180
PHI =ATAN(DTDS)*180/PI()
COH =TAU-SIGN*DTDS
```

Figure 5.5: Printout of spreadsheet cells and formulae used to calculate shear strength, instantaneous friction angle and instantaneous cohesion for a range of normal stresses.

relationship (e.g. Equation 5.5) is not expressed in terms of c and ϕ, it is necessary to devise some means for estimating the equivalent cohesive strengths and angles of friction from relationships such as those proposed by Barton.

Figure 5.4 gives definitions of the *instantaneous cohesion* c_i and the *instantaneous friction* angle ϕ_i for a normal stress of σ_n. These quantities are given by the intercept and the inclination, respectively, of the tangent to the non-linear relationship between shear strength and normal stress. These quantities may be used for stability analyses in which the Mohr-Coulomb failure criterion (Equation 5.1) is applied, provided that the normal stress σ_n is reasonably close to the value used to define the tangent point.

In a typical practical application, a spreadsheet program can be used to solve Equation 5.5 and to calculate the instantaneous cohesion and friction values for a range of normal stress values. A portion of such a spreadsheet is illustrated in Figure 5.5. Note that Equation 5.5 is not valid for $\sigma_n = 0$ and it ceases to have any practical meaning for $\phi_b + JRC \log_{10}(JCS / \sigma_n) > 70°$. This limit can be used to determine a minimum value for σ_n. An upper limit for σ_n is given by $\sigma_n = JCS$.

In the spreadsheet shown in Figure 5.5, the instantaneous friction angle ϕ_i, for a normal stress of σ_n, has been calculated from the relationship:

$$\phi_i = \arctan\left(\frac{\partial\tau}{\partial\sigma_n}\right) \tag{5.9}$$

where

$$\frac{\partial\tau}{\partial\sigma_n} = \tan\left(JRC \log_{10}\frac{JCS}{\sigma_n} + \phi_b\right) - \frac{\pi\, JRC}{\ln 10}\left[\tan^2\left(JRC \log_{10}\frac{JCS}{\sigma_n} + \phi_b\right) + 1\right] \tag{5.10}$$

The instantaneous cohesion c_i is calculated from:

$$c_i = \tau - \sigma_n \tan\phi_i \tag{5.11}$$

In choosing the values of c_i and ϕ_i for use in a particular application, the average normal stress σ_n acting on the discontinuity planes should be estimated and used to determine the appropriate row in the spreadsheet. For many practical problems in the field, a single average value of σ_n will suffice but, where critical stability problems are being considered, this selection should be made for each important discontinuity surface.

6 Analysis of structurally controlled instability

6.1 Introduction

In mining openings excavated in jointed rock masses at relatively shallow depth, the most common types of failure are those involving wedges falling from the roof or sliding out of the sidewalls of the openings. These wedges are formed by intersecting structural features, such as bedding planes and joints, which separate the rock mass into discrete but interlocked pieces. When a free face is created by the excavation of the opening, the restraint from the surrounding rock is removed. One or more of these wedges can fall or slide from the surface if the bounding planes are continuous or rock bridges along the discontinuities are broken.

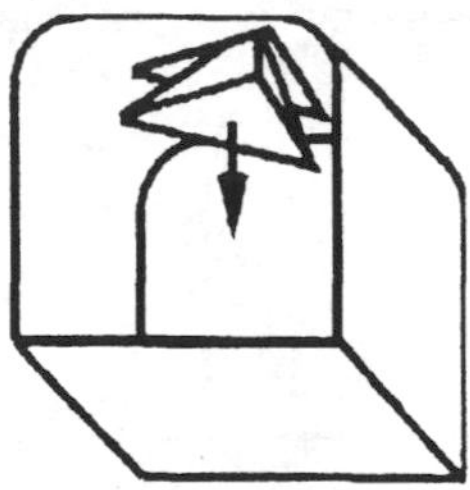

Falling wedge

Unless steps are taken to support these loose wedges, the stability of the back and walls of the opening may deteriorate rapidly. Each wedge, which is allowed to fall or slide, will cause a reduction in the restraint and the interlocking of the rock mass and this, in turn, will allow other wedges to fall. This failure process will continue until natural arching in the rock mass prevents further unravelling or until the opening is full of fallen material.

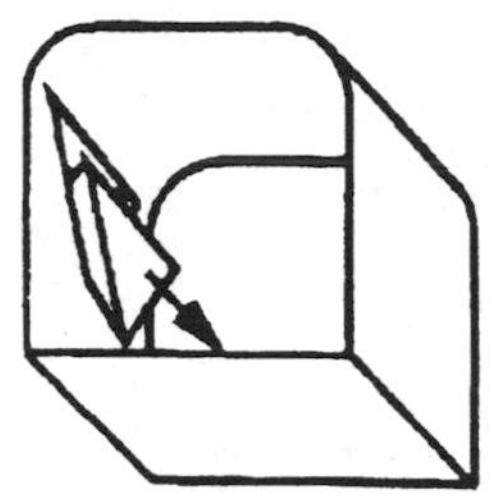

Sliding wedge

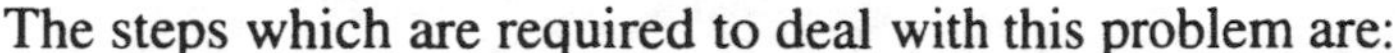

The steps which are required to deal with this problem are:

1. Determination of average dip and dip direction of significant discontinuity sets in the rock mass, as described in Chapter 3.
2. Identification of potential wedges which can slide or fall from the back or walls of the opening.
3. Calculation of the factor of safety of these wedges, depending upon the mode of failure.
4. Calculation of the amount of reinforcement required to bring the factor of safety of individual wedges up to an acceptable level.

6.2 Identification of potential wedges

The size and shape of potential wedges in the rock mass surrounding an opening depends upon the size, shape and orientation of the opening and also upon the orientation of the significant discontinuity sets. The three-dimensional geometry of the problem necessitates a set of relatively tedious calculations. While these can be performed by hand, it is far more efficient to utilise one of the computer programs which are available. One such program, called UNWEDGE[1], was developed specifically for use in underground hard rock mining and is utilised in the following discussion.

[1] This program is available from Rock Engineering Group, 12 Selwood Avenue, Toronto, Ontario, Canada M4E 1B2, Fax 1 416 698 0908, Phone 1 416 698 8217. (See order form at the end of this book).

Consider a rock mass in which three strongly developed joint sets occur. The average dips and dip directions of these sets, shown as great circles in Figure 6.1, are as follows:

Joint set	dip°	dip direction°
J1	70 ± 5	036 ± 12
J2	85 ± 8	144 ± 10
J3	55 ± 6	262 ± 15

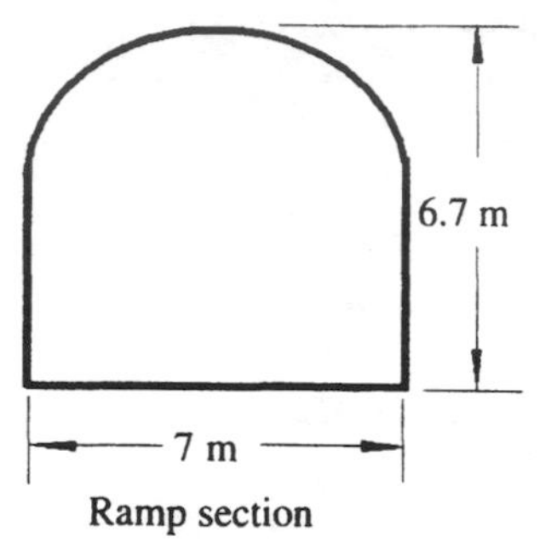

Ramp section

It is assumed that all of these discontinuities are planar and continuous and that the shear strength of the surfaces can be represented by a friction angle $\phi = 30°$ and a cohesive strength of zero. These shear strength properties are very conservative estimates, but they provide a reasonable starting point for most analyses of this type. A more detailed discussion on the shear strength of discontinuities is given in Chapter 5.

A ramp is to be excavated in this rock mass and the cross-section of the ramp is given in the margin sketch. The axis of the ramp is inclined at 15° to the horizontal or, to use the terminology associated with structural geology analysis, the ramp axis *plunges* at 15°. In the portion of the ramp under consideration in this example, the axis runs at 25° east of north or the *trend* of the axis is 025°.

The ramp axis is shown as a chain dotted line in the stereonet in Figure 6.1. The trend of the axis is shown as 025°, measured clock-

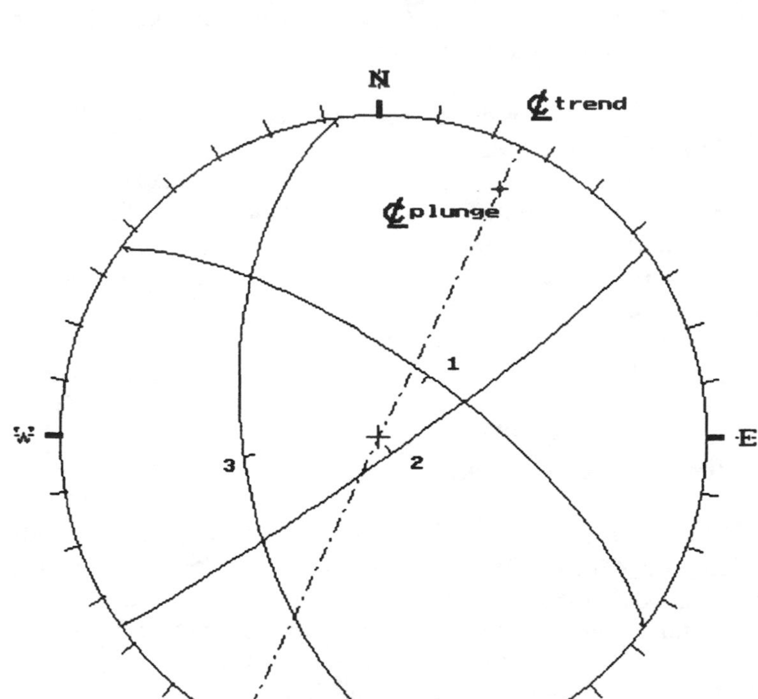

Figure 6.1: An equal area lower hemisphere plot of great circles representing the average dip and dip directions of three discontinuity sets in a rock mass. Also shown, as a chain dotted line, is the trend of the axis of a ramp excavated in this rock mass. The ramp plunge is marked with a cross.

wise from north. The plunge of the axis is 15° and this is shown as a cross on the chain dotted line representing the axis. The angle is measured inwards from the perimeter of the stereonet since this perimeter represents a horizontal reference plane.

The three structural discontinuity sets, represented by the great circles plotted in Figure 6.1, are entered into the program UNWEDGE, together with the cross-section of the ramp and the plunge and trend of the ramp axis. The program then determines the location and dimensions of the largest wedges which can be formed in the roof, floor and sidewalls of the excavation as shown in Figure 6.2.

The maximum number of simple tetrahedral wedges which can be formed by three discontinuities in the rock mass surrounding a circular tunnel is 6. In the case of a square or rectangular tunnel this number is reduced to 4. For the ramp under consideration in this example, the arched roof allows an additional wedge to form, giving a total of five. However, this additional wedge is very small and is ignored in the analysis which follows.

Note that these wedges are the largest wedges which can be formed for the given geometrical conditions. The calculation used to determine these wedges assumes that the discontinuities are ubiquitous, in other words, they can occur anywhere in the rock mass. The joints, bedding planes and other structural features included in the analysis are also assumed to be planar and continuous. These conditions mean that the analysis will always find the largest possible wedges which can form. This result can generally be considered conservative since the size of wedges, formed in actual rock masses, will be limited by the persistence and the spacing of the structural features. The program UNWEDGE allows wedges to be scaled down to more realistic sizes if it is considered that maximum wedges are unlikely to form.

Details of the four wedges illustrated in Figure 6.2 are given in the following table:

Wedge	Weight-tonnes	Failure mode	Factor of safety
Roof wedge	13	Falls	0
Side wedge 1	3.7	Slides on J1/J2	0.36
Side wedge 2	3.7	Slides on J3	0.52
Floor wedge	43	Stable	∞

The roof wedge will fall as a result of gravity loading and, because of its shape, there is no restraint from the three bounding discontinuities. This means that the factor of safety of the wedge, once it is released by excavation of the ramp opening, is zero. In some cases, sliding on one plane or along the line of intersection of two planes may occur in a roof wedge and this will result in a finite value for the factor of safety.

The two sidewall wedges are 'cousin' images of one another in that they are precisely the same shape but disposed differently in space. Consequently, the weights of these wedges are identical. The factors of safety are different since, as shown in the table, sliding occurs on different surfaces in the two cases.

The floor wedge is completely stable and requires no further consideration.

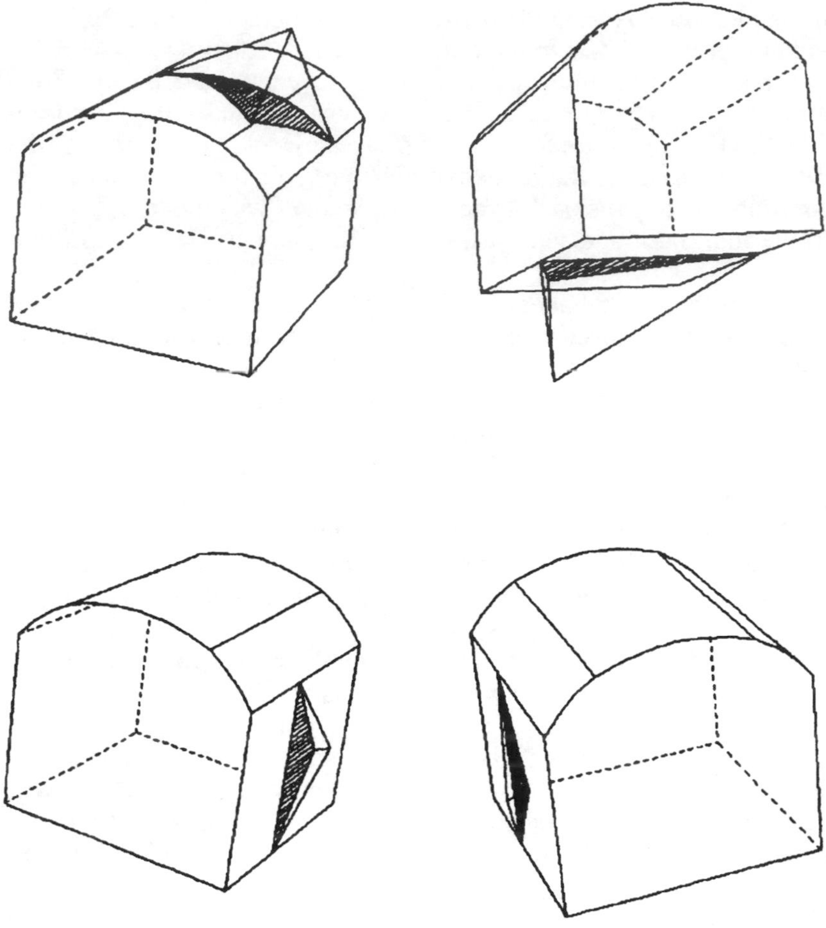

Figure 6.2: Wedges formed in the roof, floor and sidewalls of a ramp excavated in a jointed rock mass, in which the average dip and dip direction of three dominant structural features are defined by the great circles plotted in Figure 6.1.

The program UNWEDGE is intended for use in situations where the in situ stresses are low and where their influence can be neglected without the introduction of significant errors. These are the conditions in which wedge failures are most prevalent in hard rock masses.

Where high in situ stress levels occur in blocky rock masses, the factors of safety predicted by the program UNWEDGE can be incorrect. In the case of tall thin wedges, the in situ stresses will tend to clamp the wedges in place and the calculated factor of safety will be too low. On the other hand, for shallow flat wedges, the calculated factor of safety may be too high since the high in situ stresses may force the wedge out. For most practical mining situations these errors are not significant and can be compensated for by an adjustment of the factor of safety. For research into failure mechanisms and for some site applications in which the influence of in situ stresses is critical, for example large caverns, a more sophisticated method of analysis may be required.

6.3 Support to control wedge failure

A characteristic feature of wedge failures in blocky rock is that very little movement occurs in the rock mass before failure of the wedge. In the case of a roof wedge which falls, failure can occur as soon as the base of the wedge is fully exposed by excavation of the opening. For sidewall wedges, sliding of a few millimetres along one plane or the line of intersection of two planes is generally sufficient to overcome the peak strength of these surfaces. This dictates that movement along the surfaces must be minimised. Consequently, the support system has to provide a 'stiff' response to movement. This means that mechanically anchored rockbolts need to be tensioned while fully grouted rockbolts or other continuously coupled devices can be left untensioned.

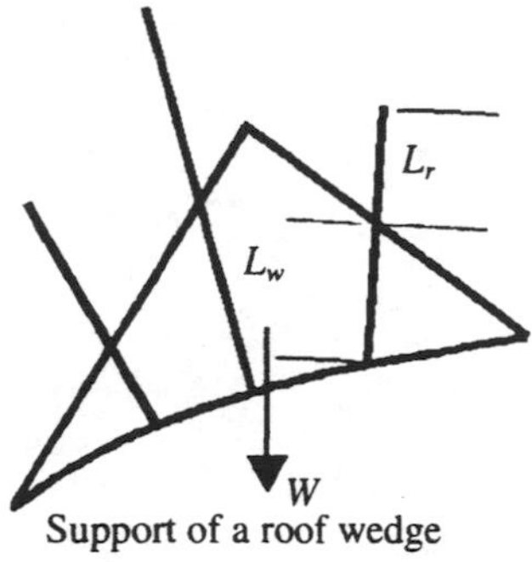

Support of a roof wedge

6.3.1 *Rock bolting wedges*

For roof wedges the total force, which should be applied by the reinforcement, should be sufficient to support the full dead weight of the wedge, plus an allowance for errors and poor quality installation. Hence, for the roof wedge illustrated in the margin sketch, the total tension applied to the rock bolts or cables should be 1.3 to 1.5 × W, giving factors of safety of 1.3 to 1.5. The lower factor of safety would be acceptable in a temporary mine access opening, such as a drilling drive, while the higher factor of safety would be used in a more permanent access opening such as a ramp.

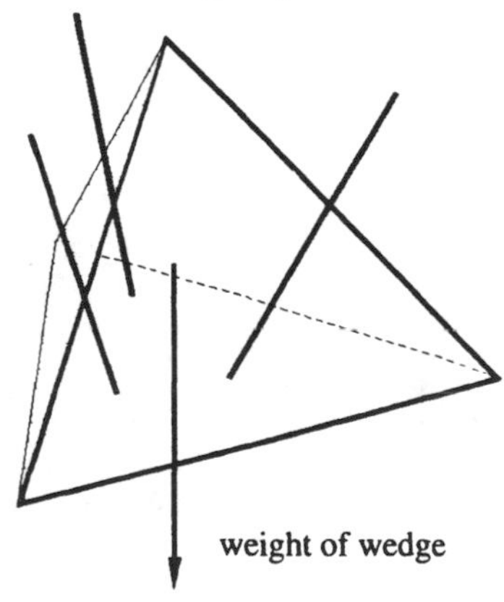

When the wedge is clearly identifiable, some attempt should be made to distribute the support elements uniformly about the wedge centroid. This will prevent any rotations which can reduce the factor of safety.

In selecting the rock bolts or cable bolts to be used, attention must be paid to the length and location of these bolts. For grouted cable bolts, the length L_w through the wedge and the length L_r in the rock behind the wedge should both be sufficient to ensure that adequate anchorage is available, as shown in the margin sketch. In the case of correctly grouted bolts or cables, these lengths should generally be about one metre. Where there is uncertainty about the quality of the grout, longer anchorage lengths should be used. When mechanically anchored bolts with face plates are used, the lengths should be sufficient to ensure that enough rock is available to distribute the loads from these attachments. These conditions are automatically checked in the program UNWEDGE.

In the case of sidewall wedges, the bolts or cables can be placed in such a way that the shear strength of the sliding surfaces is increased. As illustrated in the margin sketch, this means that more bolts or cables are placed to cross the sliding planes than across the separation planes. Where possible, these bolts or cables should be inclined so that the angle θ is between 15° and 30° since this inclination will induce the highest shear resistance along the sliding surfaces.

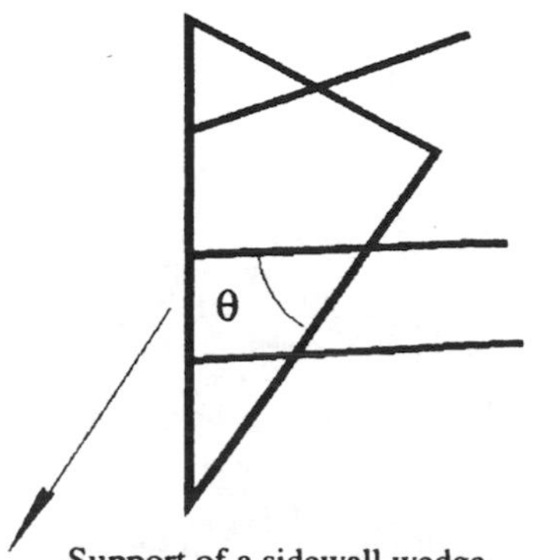

Support of a sidewall wedge

The program UNWEDGE includes a number of options for designing support for underground excavations. These include: pattern bolting, from a selected drilling position or placed normal to

Figure 6.3: Failure of a wedge where the rockbolt support was inadequate.

the excavation surface; and spot bolting, in which the location and length of the bolts are decided by the user for each installation. Mechanically anchored bolts with face plates or fully grouted bolts or cables can be selected to provide support. In addition, a layer of shotcrete can be applied to the excavation surface.

Figure 6.4 shows the rock bolt designs for the roof wedge and one of the sidewall wedges for the ramp excavation example discussed earlier. For the roof wedge, three 10 tonne capacity mechanically anchored rock bolts, each approximately 3 m long, produce a factor of safety of 1.63. The sidewall wedge, which only weighs 3.7 tonnes, requires only a single 10 tonne rock bolt for a factor of safety of 4.7. The position of the collar end of the bolt should be located for ease of drilling.

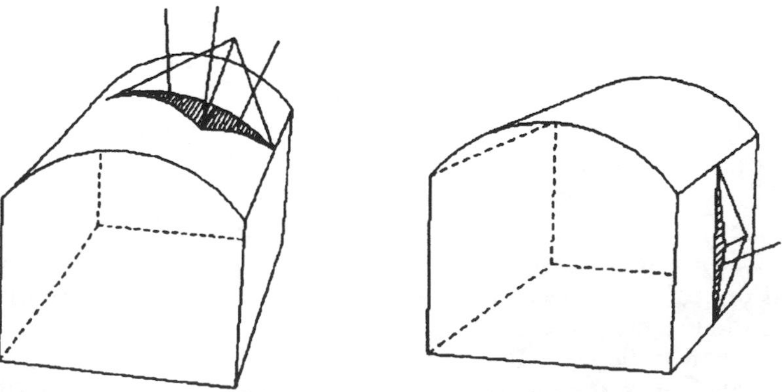

Figure 6.4: Rock bolting design for the roof wedge and one of the sidewall wedges in the ramp example discussed earlier.

6.3.2 *Shotcrete support for wedges*

Shotcrete can be used for additional support of wedges in blocky ground, and can be very effective if applied correctly. This is because the base of a typical wedge has a large perimeter and hence, even for a relatively thin layer of shotcrete, a significant cross-sectional area of the material has to be punched through before the wedge can fail.

Consider the example illustrated in Figure 6.2. The base of the roof wedge (shown cross-hatched in the upper left hand diagram) has a perimeter of 16.4 m. A layer of shotcrete 50 mm thick will mean that a total cross-sectional area of 0.8 m^2 is available to provide support for the wedge. Assuming a relatively modest shear strength for the shotcrete layer of 2 MPa (200 tonnes/m^2) means that a wedge weighing 164 tonnes can be supported. In the case of the ramp excavation discussed earlier, the wedge weighs 13 tonnes and hence a 50 mm thick layer of shotcrete would give a high ultimate factor of safety.

It is important to ensure that the shotcrete is well bonded to the rock surface in order to prevent a reduction in support capacity by peeling-off of the shotcrete layer. Good adhesion to the rock is achieved by washing the rock surface, using water only as feed to the shotcrete machine, before the shotcrete is applied.

The difficulty in using shotcrete for the support of wedges is that it has very little strength at the time of application and a period of several days is required before its full strength can be relied upon. Since wedges require immediate support, the use of shotcrete for short term stabilisation is clearly inappropriate. However, if a minimal number of rock bolts are placed to ensure that the short term stability of the rock mass is taken care of, a layer of shotcrete will provide additional long term security.

In very strong rock with large wedges, the use of shotcrete is wasteful since only that shotcrete covering the perimeter of the

Figure 6.5: Ravelling of small wedges in a closely jointed rock mass. Shotcrete can provide very effective support for such rock masses.

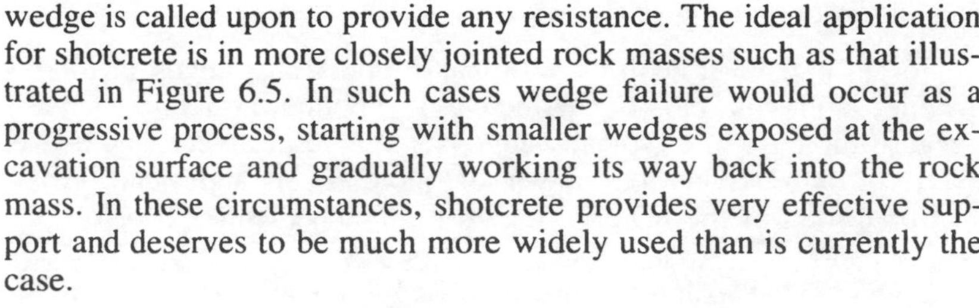
wedge is called upon to provide any resistance. The ideal application for shotcrete is in more closely jointed rock masses such as that illustrated in Figure 6.5. In such cases wedge failure would occur as a progressive process, starting with smaller wedges exposed at the excavation surface and gradually working its way back into the rock mass. In these circumstances, shotcrete provides very effective support and deserves to be much more widely used than is currently the case.

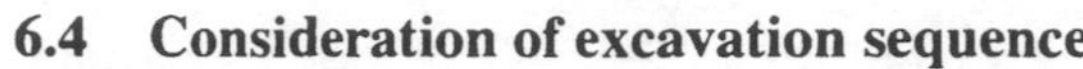
6.4 Consideration of excavation sequence

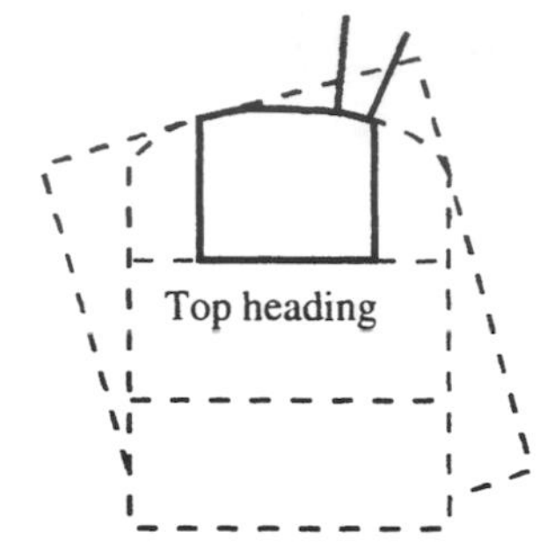

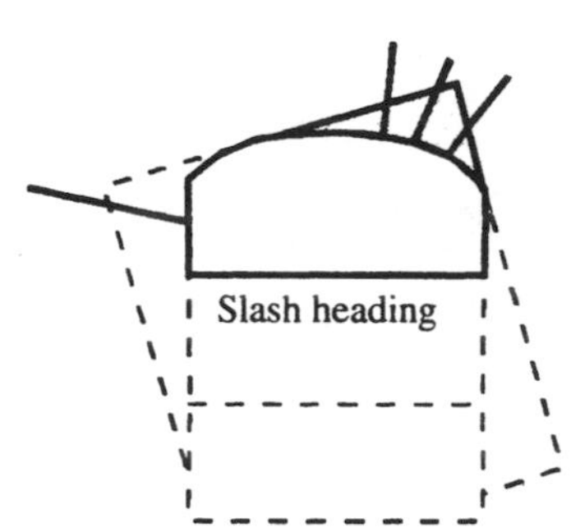

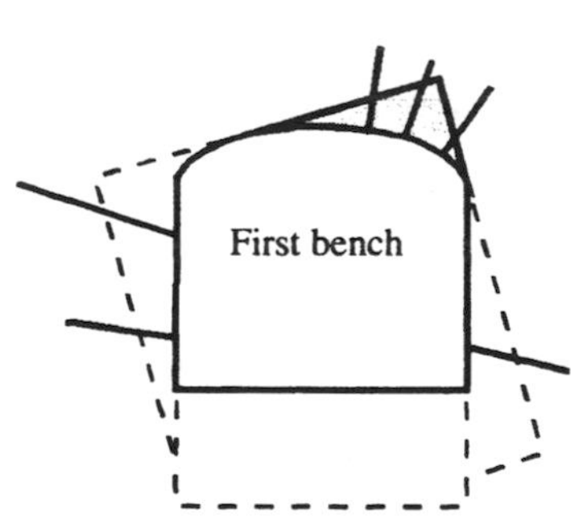

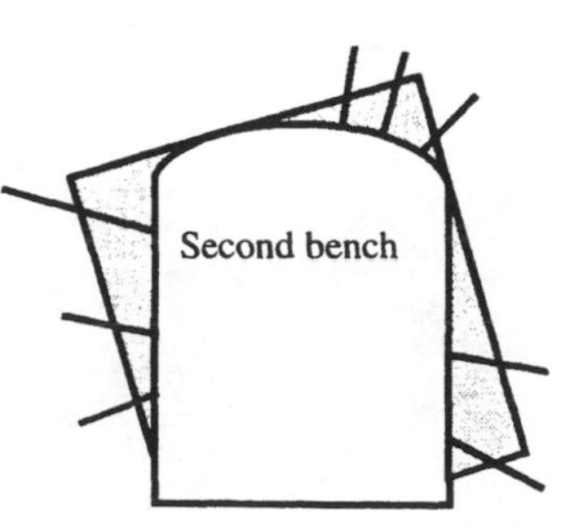

As has been emphasised several times in this chapter, wedges tend to fall or slide as soon as they are fully exposed in an excavated face. Consequently, they require immediate support in order to ensure stability. Placing this support is an important practical question to be addressed when working in blocky ground, which is prone to wedge failure.

When the structural geology of the rock mass is reasonably well understood the program UNWEDGE can be used to investigate potential wedge sizes and locations. A support pattern, which will secure these wedges, can then be designed and rockbolts can be installed as excavation progresses.

When dealing with larger excavations such as open stopes, underground crusher chambers or shaft stations, the problem of sequential support installation is a little simpler, since these excavations are usually excavated in stages. Typically, in an underground crusher chamber, the excavation is started with a top heading which is then slashed out before the remainder of the cavern is excavated by benching.

The margin sketch shows a large opening excavated in four stages with rock bolts or cables installed at each stage to support wedges, which are progressively exposed in the roof and sidewalls of the excavation. The length, orientation and spacing of the bolts or cables are chosen to ensure that each wedge is adequately supported before it is fully exposed in the excavation surface.

When dealing with large excavations of this type, the structural geology of the surrounding rock mass will have been defined from core drilling or access adits and a reasonable projection of potential wedges will be available. These projections can be confirmed by additional mapping as each stage of the excavation is completed. The program UNWEDGE provides an effective tool for exploring the size and shape of potential wedges and the support required to stabilise them.

The margin sketch shows a situation in which the support design is based upon the largest possible wedges which can occur in the roof and walls of the excavation. These wedges can sometimes form in rock masses with very persistent discontinuity surfaces such as bedding planes in layered sedimentary rocks. In many metamorphic or igneous rocks, the discontinuity surfaces are not continuous and the size of the wedges which can form is limited by the persistence of these surfaces.

The program UNWEDGE provides several options for sizing wedges. One of the most commonly measured lengths in structural mapping is the length of a joint trace on an excavation surface and one of the sizing options is based upon this trace length. The surface area of the base of the wedge, the volume of the wedge and the apex height of the wedge are all calculated by the program and all of these values can be edited by the user to set a scale for the wedge. This scaling option is very important when using the program interactively for designing support for large openings, where the maximum wedge sizes become obvious as the excavation progresses.

6.5 Application of probability theory

The program UNWEDGE has been designed for the analysis of a single wedge defined by three intersecting discontinuities. While this is adequate for many practical applications, it does not provide any facilities for selecting the three most critical joints in a large discontinuity population nor for analysing the number and location of wedges, which can form along the length of an opening such as a drive.

Early attempts have been made by a number of authors, including Tyler et al. (1991) and Hatzor and Goodman (1992), to apply probability theory to these problems and some promising results have been obtained. The analyses developed thus far are not easy to use and cannot be considered as design tools. However, these studies have shown the way for future development of such tools and it is anticipated that powerful and user-friendly methods of probabilistic analysis will be available within a few years.

7 In situ and induced stresses

7.1 Introduction

Rock at depth is subjected to stresses resulting from the weight of the overlying strata and from locked in stresses of tectonic origin. When a mine opening is excavated in this rock, the stress field is locally disrupted and a new set of stresses are induced in the rock surrounding the opening. A knowledge of the magnitudes and directions of these in situ and induced stresses is an essential component of underground excavation design since, in many cases, the strength of the rock is exceeded and the resulting instability can have serious consequences on the behaviour of the mine openings.

This chapter deals with the question of in situ stresses and also with the stress changes which are induced when mine openings are excavated in stressed rock. Problems, associated with failure of the rock around underground openings and with the design of support for these openings, will be dealt with in later chapters.

The presentation, which follows, is intended to cover only those topics which are essential for the reader to know about when dealing with the analysis of stress induced instability and the design of support to stabilise the rock under these conditions.

7.2 In situ stresses

Consider an element of rock at a depth of 1,000 m below the surface. The weight of the vertical column of rock resting on this element is the product of the depth and the unit weight of the overlying rock mass (typically about 2.7 tonnes/m^3 or 0.027 MN/m^3). Hence the vertical stress on the element is 2,700 tonnes/m^2 or 27 MPa. This stress is estimated from the simple relationship:

$$\sigma_v = \gamma z \tag{7.1}$$

where σ_v is the vertical stress
γ is the unit weight of the overlying rock and
z is the depth below surface.

Measurements of vertical stress at various mining and civil engineering sites around the world confirm that this relationship is valid although, as illustrated in Figure 7.1, there is a significant amount of scatter in the measurements.

The horizontal stresses acting on an element of rock at a depth z below the surface are much more difficult to estimate than the vertical stresses. Normally, the ratio of the average horizontal stress to the vertical stress is denoted by the letter k such that:

$$\sigma_h = k\sigma_v = k\gamma z \tag{7.2}$$

Terzaghi and Richart (1952) suggested that, for a gravitationally loaded rock mass in which no lateral strain was permitted during formation of the overlying strata, the value of k is independent of depth and is given by $k = \nu/(1 - \nu)$, where ν is the Poisson's ratio of the rock mass. This relationship was widely used in the early days of rock mechanics but, as discussed below, it proved to be inaccurate and is seldom used today.

Measurements of horizontal stresses at civil and mining sites around the world show that the ratio k tends to be high at shallow depth and that it decreases at depth (Brown and Hoek, 1978, Herget, 1988). In order to understand the reason for these horizontal stress variations it is necessary to consider the problem on a much larger scale than that of a single mine site.

Sheorey (1994) developed an elasto-static thermal stress model of the earth. This model considers curvature of the crust and variation of elastic constants, density and thermal expansion coefficients through the crust and mantle. A detailed discussion on Sheorey's model is beyond the scope of this chapter, but he did provide a simplified equation which can be used for estimating the horizontal to vertical stress ratio k. This equation is:

$$k = 0.25 + 7E_h\left(0.001 + \frac{1}{z}\right) \tag{7.3}$$

where z (m) is the depth below surface and E_h (GPa) is the average deformation modulus of the upper part of the earth's crust measured in a horizontal direction. This direction of measurement is important particularly in layered sedimentary rocks, in which the deformation modulus may be significantly different in different directions.

A plot of this equation is given in Figure 7.2 for a range of deformation moduli. The curves relating k with depth below surface z are similar to those published by Brown and Hoek (1978), Herget (1988) and others for measured in situ stresses. Hence equation 7.3 is considered to provide a sound basis for estimating the value of k.

As pointed out by Sheorey, his work does not explain the occurrence of measured vertical stresses, which are higher than the calculated overburden pressure, the presence of very high horizontal stresses at some locations or why the two horizontal stresses are seldom equal. These differences are probably due to local topographic and geological features, which cannot be taken into account in a large scale model such as that proposed by Sheorey. Consequently, where sensitivity studies have shown that the in situ stresses are likely to have a significant influence on the behaviour of underground openings, it is recommended that the in situ stresses should be measured. Suggestions for setting up a stress measuring programme are discussed later in this chapter.

7.2.1 *The World Stress Map*

The World Stress Map project, completed in July 1992, involved over 30 scientists from 18 countries and was carried out under the auspices of the International Lithosphere Project (Zoback, 1992). The aim of the project was to compile a global database of contemporary tectonic stress data. Currently over 7,300 stress orientation entries are included in a digital database. Of these approximately

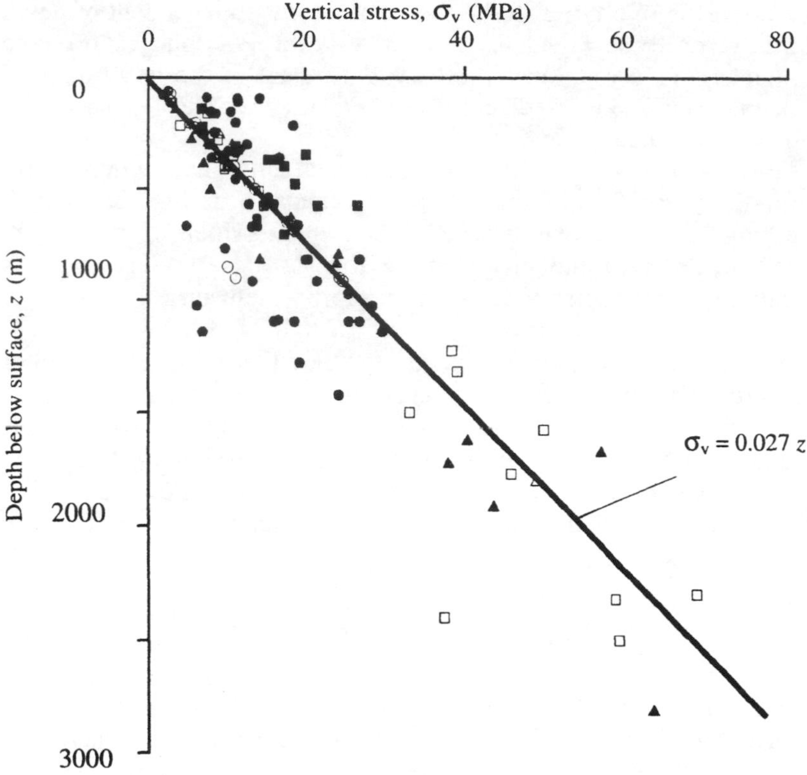

Figure 7.1: Vertical stress measurements from mining and civil engineering projects around the world (after Brown and Hoek, 1978).

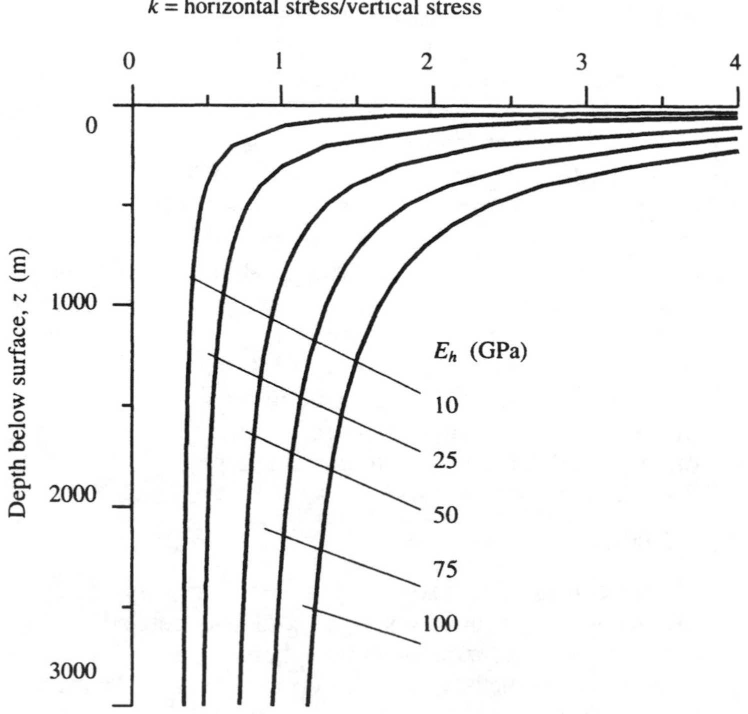

Figure 7.2: Ratio of horizontal to vertical stress for different moduli based upon Sheorey's equation. After Sheorey (1994).

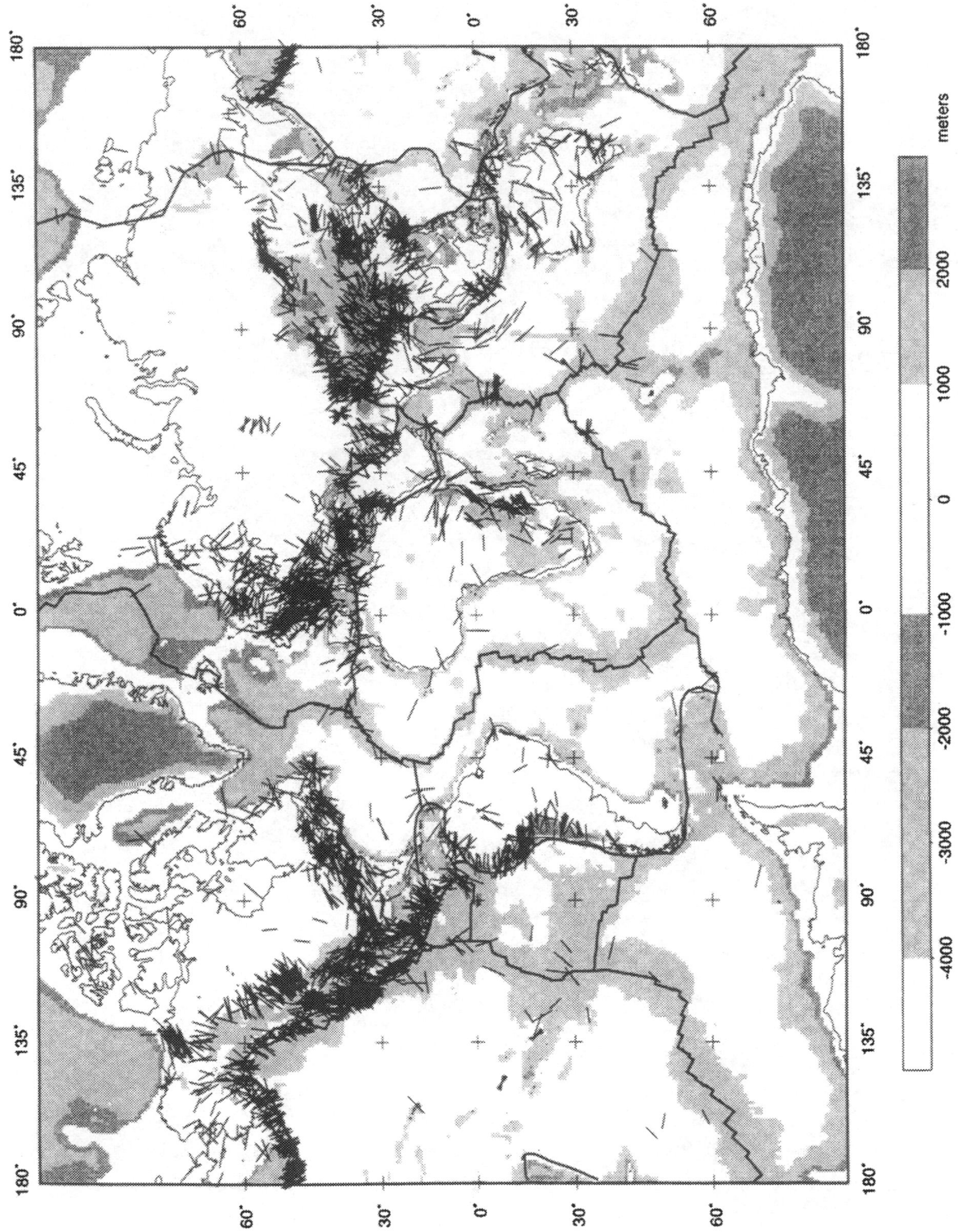

Figure 7.3: World stress map giving maximum horizontal stress orientations on a base of average topography (indicated by the shading defined in the vertical bar on the right hand side of the picture). Map provided by Dr. M.L Zoback from a paper by Zoback (1992).

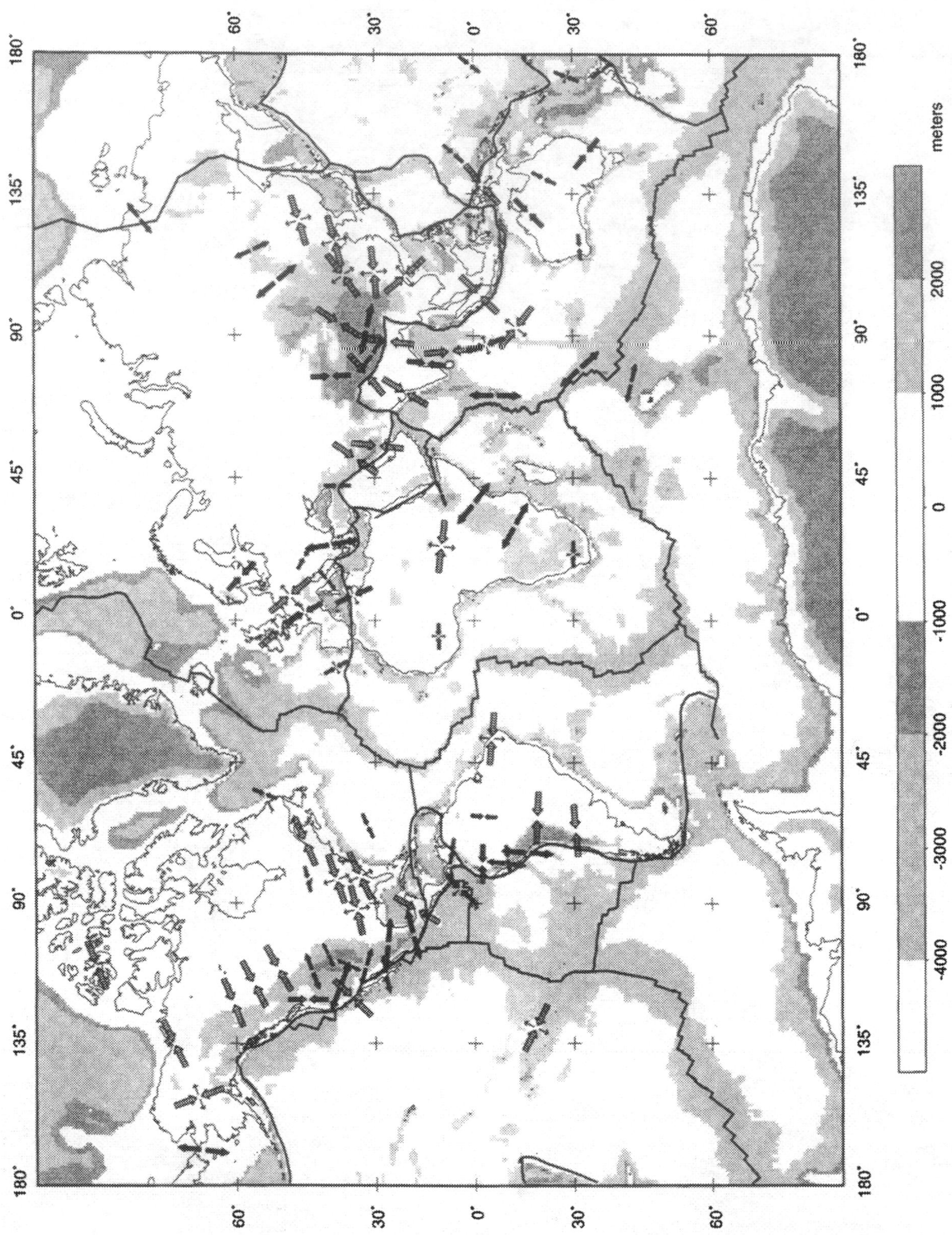

Figure 7.4: Generalised stress map showing mean directions based on average clusters of data shown in Figure 7.3. The meaning of the symbols is described in the text. Map provided by Dr M.L. Zoback from a paper by Zoback (1992).

4,400 observations are considered reliable tectonic stress indicators, recording horizontal stress orientations to within $< \pm 25°$.

The data included in the World Stress Map are derived mainly from geological observations on earthquake focal mechanisms, volcanic alignments and fault slip interpretations. Less than 5% of the data is based upon hydraulic fracturing or overcoring measurements of the type commonly used in mining and civil engineering projects.

Figure 7.3 is a version of the World Stress Map in which the orientations of maximum horizontal stress σ_{hmax} are plotted on a base of average topography. Major tectonic plate boundaries are shown as heavy lines on this map. Figure 7.4 is a generalised version of the World Stress Map which shows mean stress directions based on averages of clusters of data shown in Figure 7.3.

The stress symbols in Figure 7.4 are defined as follows:

- A single set of thick inward pointing arrows indicates σ_{hmax} orientations in a thrust faulting stress regime ($\sigma_{hmax} > \sigma_{hmin} > \sigma_v$).
- A single set of outward pointing arrows indicates σ_{hmin} orientations in a normal faulting stress regime ($\sigma_v > \sigma_{hmax} > \sigma_{hmin}$).
- Thick inward pointing arrows, indicating σ_{hmax} , together with thin outward pointing arrows, indicating σ_{hmin}, are located in strike-slip faulting stress regimes ($\sigma_{hmax} > \sigma_v > \sigma_{hmin}$).

In discussing hydraulic fracturing and overcoring stress measurements, Zoback (1992) has the following comments:

> Detailed hydraulic fracturing testing in a number of boreholes beginning very close to surface (10-20 m depth) has revealed marked changes in stress orientations and relative magnitudes with depth in the upper few hundred metres, possibly related to effects of nearby topography or a high degree of near surface fracturing.
>
> Included in the category of 'overcoring' stress measurements are a variety of stress or strain relief measurement techniques. These techniques involve a three-dimensional measurement of the strain relief in a body of rock when isolated from the surrounding rock volume; the three-dimensional stress tensor can subsequently be calculated with a knowledge of the complete compliance tensor of the rock. There are two primary drawbacks with this technique which restricts its usefulness as a tectonic stress indicator: measurements must be made near a free surface, and strain relief is determined over very small areas (a few square millimetres to square centimetres). Furthermore, near surface measurements (by far the most common) have been shown to be subject to effects of local topography, rock anisotropy, and natural fracturing (Engelder and Sbar, 1984). In addition, many of these measurements have been made for specific engineering applications (e.g. dam site evaluation, mining work), places where topography, fracturing or nearby excavations could strongly perturb the regional stress field.

Obviously, from a global or even a regional scale, the type of engineering stress measurements carried out in a mine or on a civil engineering site are not regarded as very reliable. Conversely, the World Stress Map versions presented in Figures 7.3 and 7.4 can only be used to give first order estimates of the stress directions which are likely to be encountered on a specific site. Since both stress directions and stress magnitudes are critically important in the design of underground excavations, it follows that a stress measuring programme is essential in any major underground mining or civil engineering project.

7.2.2 *Developing a stress measuring programme*

Consider the example of a new underground mine being developed at a depth of 1,000 m below surface in the Canadian Shield. The depth of the orebody is such that it is probable that in situ and induced stresses will be an important consideration in the design of the mine. Typical steps which could be followed in the analysis of this problem are:

a) During preliminary mine design, the information presented in equations 7.1, 7.2 and 7.3 can be used to obtain a first rough estimate of the vertical and average horizontal stress in the vicinity of the orebody. For a depth of 1,000 m, these equations give the vertical stress $\sigma_v = 27$ MPa , the ratio $k = 1.3$ (for $E_h = 75$ GPa) and hence the average horizontal stress $\sigma_h = 35.1$ MPa. A preliminary analysis of the stresses induced around the proposed mine stopes (as described later in this chapter) shows that these induced stresses are likely to exceed the strength of the rock and that the question of stress must be considered in more detail. Note that for many openings in strong rock at shallow depth, stress problems may not be significant and the analysis need proceed no further.

b) For this particular case, stress problems are considered to be important. A typical next step would be to search the literature in an effort to determine whether the results of in situ stress measurement programmes are available for mines or civil engineering projects within a radius of say 50 km of the site. Since this particular project is in the Canadian shield, the publications of Herget, summarised in his book *Stresses in Rock* (1988), would be a useful starting point for such a search. With luck, a few stress measurement results will be available for the region in which the new mine is located and these results can be used to refine the analysis described earlier.

c) Assuming that the results of the analysis of induced stresses in the rock surrounding the proposed stopes indicate that significant zones of rock failure are likely to develop, and that support costs are likely to be high, it is probably justifiable to set up a stress measurement project on the mine site. These measurements can be carried out in deep boreholes from the surface, using hydraulic fracturing techniques, or from underground access using overcoring methods. The choice of the method and the number of measurements to be carried out depends upon the urgency of the problem, the availability of underground access and the costs involved in the project. Note that very few mines have access to the equipment required to carry out a stress measurement project and, rather than purchase this equipment, it may be worth bringing in an organisation which has the equipment and which specialises in such measurements.

Many orebodies are associated with regional tectonic features such as major faults. Hence, the in situ stresses in the vicinity of the orebody may be rotated with respect to the regional stress field, and may be significantly different in magnitude from the values estimated from the general trends described earlier. These differences can be very important in the design of the stopes and in the selection of support and, where it is suspected that this is likely to be the case, in situ stress measurements become an essential component of the overall mine design process.

7.3 Analysis of induced stresses

When an underground opening is excavated into a stressed rock mass, the stresses in the vicinity of the new opening are re-distributed. Consider the example of the stresses induced in the rock surrounding a horizontal borehole as illustrated in Figure 7.5, showing a vertical slice normal to the borehole axis.

Before the borehole is drilled, the in situ stresses σ_v, σ_{h1} and σ_{h2} are uniformly distributed in the slice of rock under consideration. After removal of the rock from within the borehole, the stresses in the immediate vicinity of the borehole are changed and new stresses are induced. Three principal stresses σ_1, σ_2 and σ_3 acting on a typical element of rock are shown in Figure 7.5.

The convention used in rock mechanics is that *compressive* stresses are always *positive* and the three principal stresses are numbered such that σ_1 is the largest and σ_3 is the smallest (algebraically) of the three.

The three principal stresses are mutually perpendicular, but they may be inclined to the direction of the applied in situ stress. This is evident in Figure 7.6, which shows the directions of the stresses in the rock surrounding a horizontal borehole subjected to a horizontal in situ stress σ_{h1} equal to three times the vertical in situ stress σ_v. The longer bars in this figure represent the directions of the maximum principal stress σ_1, while the shorter bars give the directions of the minimum principal stress σ_3 at each element considered. In this particular case, σ_2 is coaxial with the in situ stress σ_{h2}, but the other principal stresses σ_1 and σ_3 are inclined to σ_{h1} and σ_v.

Contours of the magnitudes of the maximum principal stress σ_1 and the minimum principal stress σ_3 are given in Figure 7.7. This figure shows that the redistribution of stresses is concentrated in the rock very close to the borehole and that, at a distance of say three times the radius from the centre of the hole, the disturbance to the in situ stress field is negligible.

An analytical solution for the stress distribution in a stressed elastic plate containing a circular hole was published by Kirsch (1898) and this formed the basis for many early studies of rock behaviour around tunnels and shafts.

Following along the path pioneered by Kirsch, researchers such as Love (1927), Muskhelishvili (1953) and Savin (1961) published solutions for excavations of various shapes in elastic plates. A useful summary of these solutions and their application in rock mechanics was published by Brown in an introduction to a volume entitled *Analytical and Computational Methods in Engineering Rock Mechanics* (1987).

Closed form solutions still possess great value for conceptual understanding of behaviour and for the testing and calibration of numerical models. For design purposes, however, these models are restricted to very simple geometries and material models. They are of limited practical value.

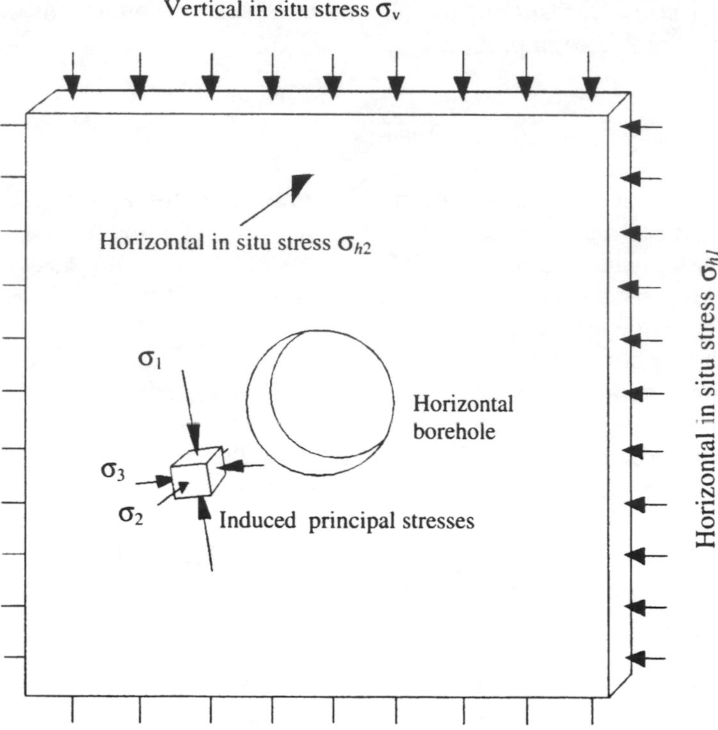

Figure 7.5: Illustration of principal stresses σ_1, σ_2 and σ_3 induced in an element of rock close to a horizontal borehole subjected to a vertical in situ stress σ_v, a horizontal in situ stress σ_{h1} in a plane normal to the borehole axis and a horizontal in situ stress σ_{h2} parallel to the borehole axis.

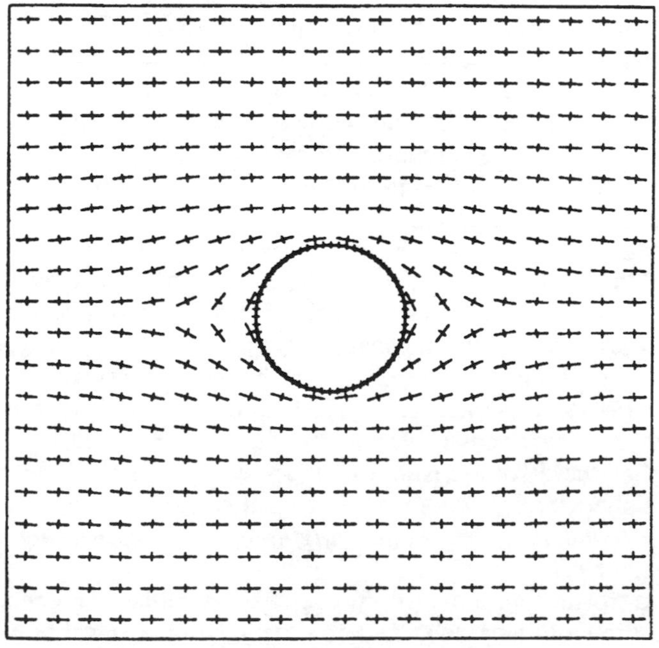

Figure 7.6: Principal stress directions in the rock surrounding a horizontal borehole subjected to a horizontal in situ stress σ_{h1} equal to 3 σ_v, where σ_v is the vertical in situ stress.

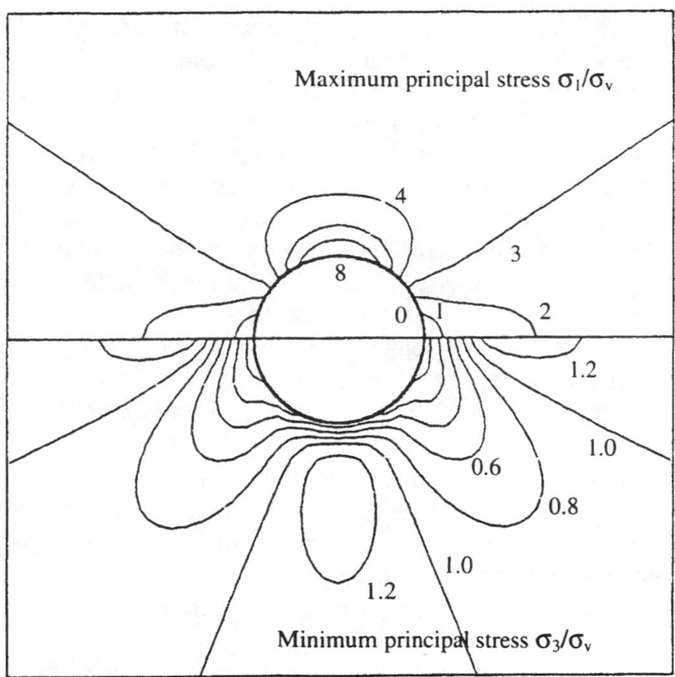

Figure 7.7: Contours of maximum and minimum principal stress magnitudes in the rock surrounding a horizontal borehole, subjected to a vertical in situ stress of σ_v and a horizontal in situ stress of $3\sigma_v$.

7.3.1 *Numerical methods of stress analysis*

Most underground mining excavations are irregular in shape and are frequently grouped close to other excavations. These groups of excavations, which may be stopes or the various service openings associated with a ramp or shaft system, form a set of complex three-dimensional shapes. In addition, since orebodies are frequently associated with geological features such as faults and intrusions, the rock properties are seldom uniform within the rock volume of interest. Consequently, the closed form solutions described earlier are of limited value in calculating the stresses, displacements and failure of the rock mass surrounding these mining excavations. Fortunately a number of computer-based numerical methods have been developed over the past few decades and these methods provide the means for obtaining approximate solutions to these problems.

Numerical methods for the analysis of stress driven problems in rock mechanics can be divided into two classes:

- *Boundary methods*, in which only the boundary of the excavation is divided into elements and the interior of the rock mass is represented mathematically as an infinite continuum.
- *Domain methods*, in which the interior of the rock mass is divided into geometrically simple elements each with assumed properties. The collective behaviour and interaction of these simplified elements model the more complex overall behaviour of the rock

mass. *Finite element* and *finite difference* methods are domain techniques which treat the rock mass as a continuum. The *distinct element* method is also a domain method which models each individual block of rock as a unique element.

These two classes of analysis can be combined in the form of *hybrid models* in order to maximise the advantages and minimise the disadvantages of each method.

It is possible to make some general observations about the two types of approaches discussed above. In domain methods, a significant amount of effort is required to create the mesh which is used to divide the rock mass into elements. In the case of complex models, such as those containing multiple openings, meshing can become extremely difficult. The availability of highly optimised mesh-generators in many models makes this task much simpler than was the case when the mesh had to be created manually. In contrast, boundary methods require only that the excavation boundary be discretized and the surrounding rock mass is treated as an infinite continuum. Since fewer elements are required in the boundary method, the demand on computer memory and on the skill and experience of the user is reduced.

In the case of domain methods, the outer boundaries of the model must be placed sufficiently far away from the excavations, that errors, arising from the interaction between these outer boundaries and the excavations, are reduced to an acceptable minimum. On the other hand, since boundary methods treat the rock mass as an infinite continuum, the far field conditions need only be specified as stresses acting on the entire rock mass and no outer boundaries are required. The main strength of boundary methods lies in the simplicity achieved by representing the rock mass as a continuum of infinite extent. It is this representation, however, that makes it difficult to incorporate variable material properties and the modelling of rock-support interaction. While techniques have been developed to allow some boundary element modelling of variable rock properties, these types of problems are more conveniently modelled by domain methods.

Before selecting the appropriate modelling technique for particular types of problems, it is necessary to understand the basic components of each technique.

Boundary Element Method

The boundary element method derives its name from the fact that only the boundaries of the problem geometry are divided into elements. In other words, only the excavation surfaces, the free surface for shallow problems, joint surfaces where joints are considered explicitly and material interfaces for multi-material problems are divided into elements. In fact, several types of boundary element models are collectively referred to as 'the boundary element method'. These models may be grouped as follows:

1. Indirect (Fictitious Stress) method, so named because the first step in the solution is to find a set of fictitious stresses which satisfy prescribed boundary conditions. These stresses are then used in the calculation of actual stresses and displacements in the rock mass.

2. Direct method, so named because the displacements are solved directly for the specified boundary conditions.
3. Displacement Discontinuity method, so named because it represents the result of an elongated slit in an elastic continuum being pulled apart.

The differences between the first two methods are not apparent to the program user. The direct method has certain advantages in terms of program development, as will be discussed later in the section on Hybrid approaches.

The fact that a boundary element model extends 'to infinity' can also be a disadvantage. For example, a heterogeneous rock mass consists of regions of finite, not infinite, extent. Special techniques must be used to handle these situations. Joints are modelled explicitly in the boundary element method using the displacement discontinuity approach, but this can result in a considerable increase in computational effort. Numerical convergence is often found to be a problem for models incorporating many joints. For these reasons, problems, requiring explicit consideration of several joints and/or sophisticated modelling of joint constitutive behaviour, are often better handled by one of the remaining numerical methods.

A widely-used application of displacement discontinuity boundary elements is in the modelling of tabular ore bodies. Here, the entire ore seam is represented as a 'discontinuity' which is initially filled with ore. Mining is simulated by reduction of the ore stiffness to zero in those areas where mining has occurred, and the resulting stress redistribution to the surrounding pillars may be examined (Salamon, 1974, von Kimmelmann et al., 1984).

Further details on boundary element methods can be found in the book *Boundary element methods in solid mechanics* by Crouch and Starfield (1983).

Finite element and finite difference methods

In practice, the finite element method is usually indistinguishable from the finite difference method; thus, they will be treated here as one and the same. For the boundary element method, it was seen that conditions on a surface could be related to the state at all points throughout the remaining rock, even to infinity. In comparison, the finite element method relates the conditions at a few points within the rock (nodal points) to the state within a finite closed region formed by these points (the element). The physical problem is modelled numerically by dividing the entire problem region into elements.

The finite element method is well suited to solving problems involving heterogeneous or non-linear material properties, since each element explicitly models the response of its contained material. However, finite elements are not well suited to modelling infinite boundaries, such as occur in underground excavation problems. One technique for handling infinite boundaries is to discretize beyond the zone of influence of the excavation and to apply appropriate boundary conditions to the outer edges. Another approach has been to develop elements for which one edge extends to infinity i.e. so-called infinity' finite elements. In practice, efficient pre- and post-processors allow the user to perform parametric analyses and assess the influence of approximated far-field boundary conditions. The time required for this process is negligible compared to the total analysis time.

Joints can be represented explicitly using specific 'joint elements'. Different techniques have been proposed for handling such elements, but no single technique has found universal favour. Joint interfaces may be modelled, using quite general constitutive relations, though possibly at increased computational expense depending on the solution technique.

Once the model has been divided into elements, material properties have been assigned and loads have been prescribed, some technique must be used to redistribute any unbalanced loads and thus determine the solution to the new equilibrium state. Available solution techniques can be broadly divided into two classes-implicit and explicit. Implicit techniques assemble systems of linear equations which are then solved using standard matrix reduction techniques. Any material non-linearity is accounted for by modifying stiffness coefficients (secant approach) and/or by adjusting prescribed variables (initial stress or initial strain approach). These changes are made in an iterative manner such that all constitutive and equilibrium equations are satisfied for the given load state.

The response of a non-linear system generally depends upon the sequence of loading. Thus it is necessary that the load path modelled be representative of the actual load path experienced by the body. This is achieved by breaking the total applied load into load increments, each increment being sufficiently small, that solution convergence for the increment is achieved after only a few iterations. However, as the system being modelled becomes increasingly non-linear and the load increment represents an ever smaller portion of the total load, the incremental solution technique becomes similar to modelling the quasi-dynamic behaviour of the body, as it responds to gradual application of the total load.

In order to overcome this, a 'dynamic relaxation' solution technique was proposed (Otter et al., 1966) and first applied to geomechanics modelling by Cundall (1971). In this technique no matrices are formed. Rather, the solution proceeds explicitly-unbalanced forces, acting at a material integration point, result in acceleration of the mass associated with the point; applying Newton's law of motion expressed as a difference equation yields incremental displacements; applying the appropriate constitutive relation produces the new set of forces, and so on marching in time, for each material integration point in the model. This solution technique has the advantage, that both geometric and material non-linearities are accommodated, with relatively little additional computational effort as compared to a corresponding linear analysis, and computational expense increases only linearly with the number of elements used. A further practical advantage lies in the fact that numerical divergence usually results in the model predicting obviously anomalous physical behaviour. Thus, even relatively inexperienced users may recognise numerical divergence.

Most commercially available finite element packages use implicit (i.e. matrix) solution techniques. For linear problems and problems of moderate non-linearity, implicit techniques tend to perform faster than explicit solution techniques. However, as the degree of non-linearity of the system increases, imposed loads must be applied in smaller increments which implies a greater number of matrix re-formations and reductions, and hence increased computational expense. Therefore, highly non-linear problems are best handled by packages using an explicit solution technique.

Distinct Element Method

In ground conditions conventionally described as blocky (i.e. where the spacing of the joints is of the same order of magnitude as the excavation dimensions), intersecting joints form wedges of rock that may be regarded as rigid bodies. That is, these individual pieces of rock may be free to rotate and translate, and the deformation, that takes place at block contacts, may be significantly greater than the deformation of the intact rock, so that individual wedges may be considered rigid. For such conditions it is usually necessary to model many joints explicitly. However, the behaviour of such systems is so highly non-linear, that even a jointed finite element code, employing an explicit solution technique, may perform relatively inefficiently.

An alternative modelling approach is to develop data structures that represent the blocky nature of the system being analysed. Each block is considered a unique free body that may interact at contact locations with surrounding blocks. Contacts may be represented by the overlaps of adjacent blocks, thereby avoiding the necessity of unique joint elements. This has the added advantage that arbitrarily large relative displacements at the contact may occur, a situation not generally tractable in finite element codes.

Due to the high degree of non-linearity of the systems being modelled, explicit solution techniques are favoured for distinct element codes. As is the case for finite element codes employing explicit solution techniques, this permits very general constitutive modelling of joint behaviour with little increase in computational effort and results in computation time being only linearly dependent on the number of elements used. The use of explicit solution techniques places fewer demands on the skills and experience than the use of codes employing implicit solution techniques.

Although the distinct element method has been used most extensively in academic environments to date, it is finding its way into the offices of consultants, mine planners and designers. Further experience in the application of this powerful modelling tool to practical design situations and subsequent documentation of these case histories is required, so that an understanding may be developed of where, when and how the distinct element method is best applied.

Hybrid approaches

The objective of a hybrid method is to combine the above methods in order to eliminate undesirable characteristics while retaining as many advantages as possible. For example, in modelling an underground excavation, most non-linearity will occur close to the excavation boundary, while the rock mass at some distance will behave in an elastic fashion. Thus, the near-field rock mass might be modelled, using a distinct element or finite element method, which is then linked at its outer limits to a boundary element model, so that the far-field boundary conditions are modelled exactly. In such an approach, the direct boundary element technique is favoured as it results in increased programming and solution efficiency.

Lorig and Brady (1984) used a hybrid model consisting of a discrete element model for the near field and a boundary element model for the far field conditions in a rock mass surrounding a circular tunnel.

7.3.2 *Two-dimensional and three-dimensional models*

A two-dimensional model, such as that illustrated in Figure 7.5, can be used for the analysis of stresses and displacements in the rock surrounding a tunnel, shaft or borehole, where the length of the opening is much larger than its cross-sectional dimensions. The stresses and displacements in a plane, normal to the axis of the opening, are not influenced by the ends of the opening, provided that these ends are far enough away.

On the other hand, a stope in an underground mine has a much more equi-dimensional shape and the effect of the end walls of the stope cannot be neglected. In this case, it is much more appropriate to carry out a three-dimensional analysis of the stresses and displacements in the surrounding rock mass. Unfortunately, this switch from two to three dimensions is not as simple as it sounds and there are relatively few good three-dimensional numerical models, which are suitable for routine stress analysis work in a typical mining environment.

EXAMINE^{3D}[1] and MAP^{3D}[2] are three-dimensional boundary element programs which provide a starting point for an analysis of a problem in which the three-dimensional geometry of the openings is important. Such a three-dimensional analysis provides a clear indication of stress concentrations and of the influence of the three-dimensional geometry of the problem. In many cases, it is possible to simplify the problem to two-dimensions by considering the stresses on critical sections identified in the three-dimensional model.

More sophisticated three-dimensional finite element models such as VISAGE[3] are available, but are not particularly easy to use at the present time. In addition, definition of the input parameters and interpretation of the results of these models would stretch the capabilities of all but the most experienced modellers. It is probably best to leave this type of modelling in the hands of these specialists.

It is recommended that, where the problem being considered is obviously three-dimensional, a preliminary elastic analysis be carried out by means of one of the three-dimensional boundary element programs. The results can then be used to decide whether further three-dimensional analyses are required or whether appropriate two-dimensional sections can be modelled using a program such as PHASES, described in the following section.

---

[1]Available from The Rock Engineering Group, 12 Selwood Avenue, Toronto, Ontario, Canada M4E 1B2, Fax 1 416 698 0908, Phone 1 416 698 8217.

[2]Available from Mine Modelling Limited, 16 Park Street, P.O. Box 386, Copper Cliff, Ontario P0M 1N0, Fax 1 705 682 0087, Phone 1 705 682 1572.

[3]Available from Vector International Processing Systems Ltd., Suites B05 and B06, Surrey House, 34 Eden Street, Kingston on Thames, KT1 1ER, England. Fax 44 81 541 4550, Phone 44 81 549 3444.

7.3.3 *Stress analysis using the program PHASES*

In order to meet the requirements of modelling the post-failure behaviour of rock masses and the interaction of these rocks with support, a two-dimensional hybrid model called PHASES[4] was developed at the University of Toronto. This program uses finite elements to model the heterogeneous non-linear behaviour of the rock close to the excavation boundaries. Far field in situ stress conditions are modelled by means of a boundary element model.

The program will be used in later chapters dealing with progressive failure and support interaction. At this stage its use will be restricted to a simple example of the elastic stress distribution in a homogeneous rock mass, surrounding two adjacent stopes, in an orebody dipping at approximately 20° to the horizontal. No provision is made for different rock types and a homogeneous material model is used for this analysis. These stopes are assumed to have a long strike length so that a two-dimensional model can be used for the stress analysis. In situ stresses are: $\sigma_v = 20$ MPa (vertical), $\sigma_{h2} = 30$ MPa (parallel to strike) and $\sigma_{h1} = 40$ MPa (normal to strike).

The principal stress directions, shown in Figure 7.8, illustrate the re-distribution of stress around the two adjacent stopes and the flow of stress resulting in a concentration of stress in the pillar. Displacement vectors in the rock mass are shown in Figure 7.9 and these indicate a significant closure of the two stopes. Note that the influence of this stope closure extends a considerable distance out into the surrounding rock mass. Figure 7.10 illustrates contours of maximum principal stress (σ_1), showing high compressive stresses in the pillar and around the outer corners of the stopes. The relaxation of stresses in the back and floor of the stopes is evident in this figure and these stress reductions can be just as important as the high compressive stress concentrations, when considering the stability of the rock mass surrounding the openings. The minimum principal stress (σ_3) contours, shown in Figure 7.11, indicate a zone of tensile stress in the rock above and below the stopes. The relaxation resulting from these tensile stresses can cause instability if the support system is inadequate.

[4]Available from The Rock Engineering Group, 12 Selwood Avenue, Toronto, Ontario, Canada M4E 1B2, Fax 1 416 698 0908, Phone 1 416 698 8217. (See order form at the end of this book).

Figure 7.8: Principal stress trajectories in the rock surrounding two adjacent stopes. The longer of the legs of each cross gives the direction of the maximum principal stress, σ_1, at each element.

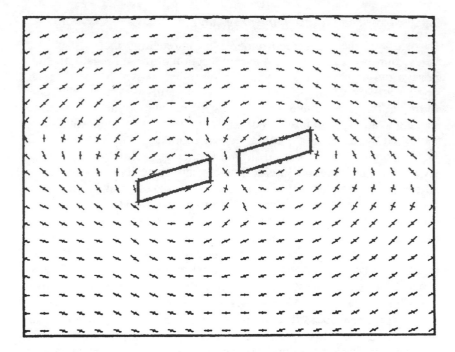

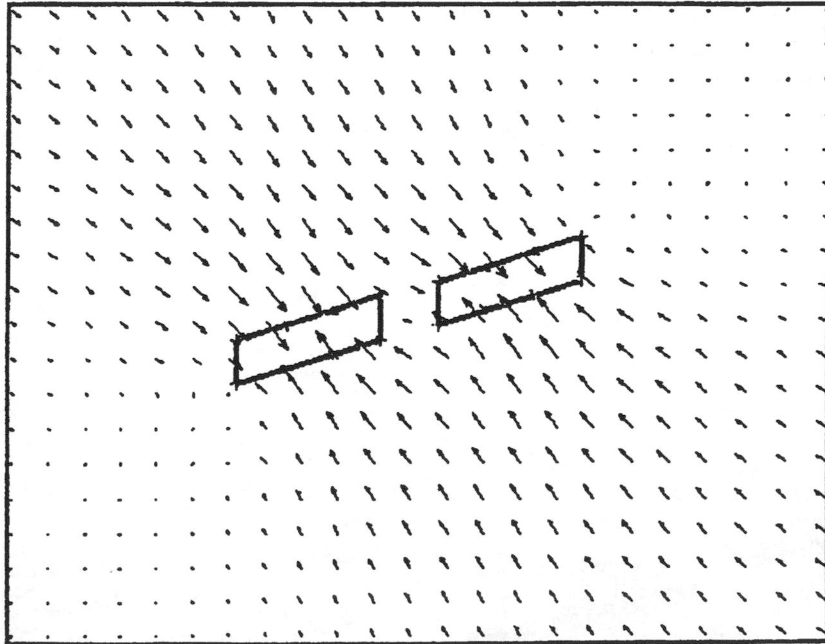

Figure 7.9: Displacement in the rock mass surrounding two adjacent stopes. The length of the arrows gives the magnitude of the displacement at each element.

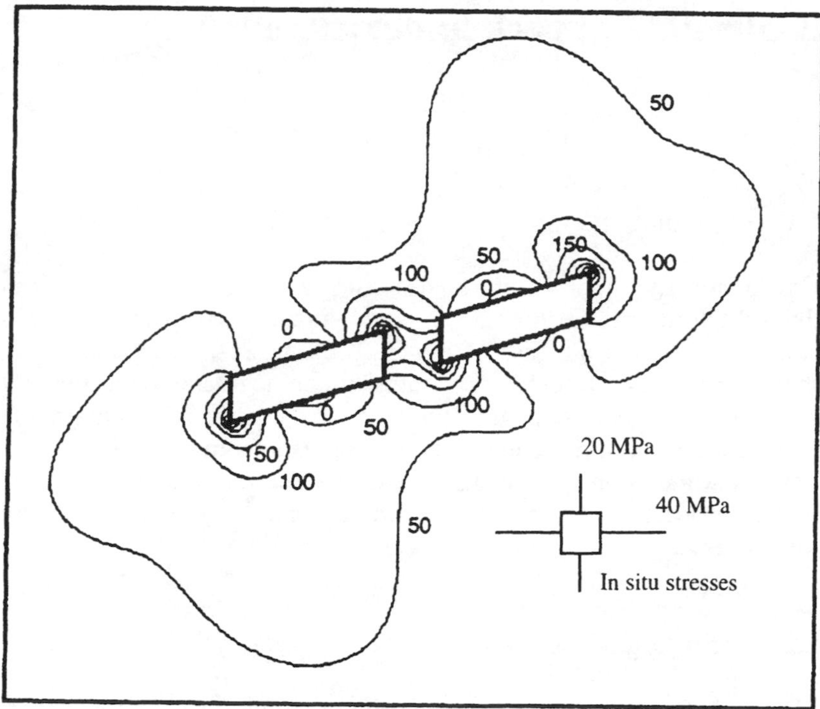

Figure 7.10 : Contours of maximum principal stress (σ_1) in the rock mass surrounding two adjacent stopes.

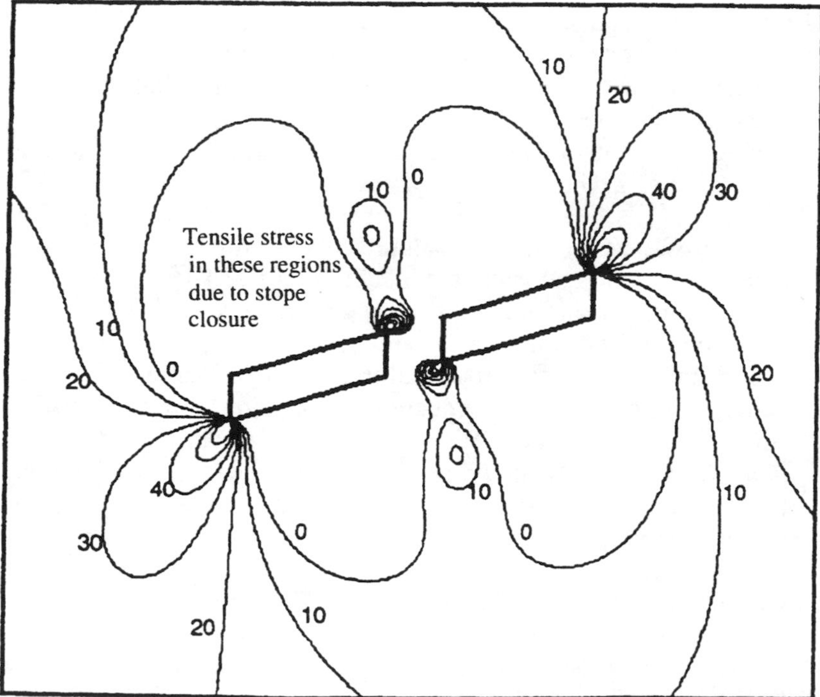

Figure 7.11: Minimum principal stress (σ_3) contours in the rock mass surrounding two adjacent stopes.

8 Strength of rock and rock masses

8.1 Introduction

One of the major problems in designing underground openings is that of estimating the strength and deformation properties of the in situ rock mass. In the case of jointed rock masses, an evaluation of these properties presents formidable theoretical and experimental problems. However, since this question is of fundamental importance in almost all major designs involving excavations in rock, it is essential that some attempt be made to estimate these strength and deformation properties and that these estimates should be as realistic and reliable as possible.

8.2 Definition of the problem

Table 8.1 illustrates the range of problems to be considered. Understanding the behaviour of jointed rock masses requires a study of the intact rock material and of the individual discontinuity surfaces which go together to make up the system. Depending upon the number, orientation and nature of the discontinuities, the intact rock pieces will translate, rotate or crush in response to stresses imposed upon the rock mass. Since a large number of possible combinations of block shapes and sizes exist, it is obviously necessary to find any behavioural trends which are common to all of these combinations. The establishment of such common trends is the most important objective in this chapter.

Before embarking upon a study of the individual components and of the system as a whole, it is necessary to set down some basic definitions.

- *Intact rock* refers to the unfractured blocks which occur between structural discontinuities in a typical rock mass. These pieces may range from a few millimetres to several metres in size and their behaviour is generally elastic and isotropic. For most hard igneous and metamorphic rocks failure can be classified as brittle, which implies a sudden reduction in strength when a limiting stress level is exceeded. Weak sedimentary rocks may fail in a more ductile manner, in which there is little or no strength reduction when a limiting stress level is reached. Viscoelastic or time-dependent behaviour is not usually considered to be significant unless one is dealing with evaporites such as salt or potash. The mechanical properties of these viscoelastic materials are not dealt with in this volume.
- *Joints* are a particular type of geological discontinuity, but the term tends to be used generically in rock mechanics and it usually covers all types of structural weakness. The shear strength of such structural weakness planes is discussed in Chapter 5.
- *Strength*, in the context of this discussion, refers to the maximum

stress level which can be carried by a specimen. The presentation of rock strength data and its incorporation into a failure criterion depends upon the preference of the individual and upon the end use for which the criterion is intended. In dealing with gravity driven wedge failure problems, where limit equilibrium methods of analyses are used, the most useful failure criterion is one which expresses the shear strength in terms of the effective normal stress, acting across a particular weakness plane or shear zone, as discussed in Chapter 5. On the other hand, when analysing the stability of underground excavations in medium to high stress regimes, the response of the rock to the principal stresses acting upon each element is of paramount interest. Consequently, for the underground excavation engineer, a plot of triaxial test data, in terms of the major principal stress at failure versus minimum principal stress, is the most useful form of failure criterion.

8.3 Strength of intact rock

A vast amount of information on the strength of intact rock has been published during the past fifty years and it would be inappropriate to attempt to review all this information here. Interested readers are referred to the excellent review presented by Jaeger (1971).

Hoek and Brown (1980a, 1980b) and Hoek (1983) reviewed the published information on intact rock strength and proposed an empirical failure criterion for rock. In developing their empirical failure criterion, Hoek and Brown attempted to satisfy the following conditions:

a) The failure criterion should give good agreement with rock strength values determined from laboratory triaxial tests on core samples of intact rock. These samples are typically 50 mm in diameter and should be oriented perpendicular to any discontinuity surfaces in the rock.
b) The failure criterion should be expressed by mathematically simple equations based, to the maximum extent possible, upon dimensionless parameters.
c) The failure criterion should offer the possibility of extension to deal with the failure of jointed rock masses.

Based on their experimental and theoretical experience with the fracture mechanics of rock, Hoek and Brown (1980a, 1980b) experimented with a number of distorted parabolic curves to find one which gave good coincidence with the original Griffith theory (Griffith, 1921, 1924). Griffith was concerned with brittle failure in glass and he expressed his relationship in terms of tensile stresses. Hoek and Brown sought a relationship which fitted the observed failure conditions for brittle rocks subjected to compressive stress conditions.

Note that the process used by Hoek and Brown in deriving their empirical failure criterion was one of pure trial and error. Apart from the conceptual starting point provided by the Griffith theory, there is no fundamental relationship between the empirical constants included in the criterion and any physical characteristics of the rock. The justification for choosing this particular criterion over the

Table 8.1: Summary of rock mass characteristics, testing methods and theoretical considerations.

	Description	Strength characteristics	Strength testing considerations	Theoretical
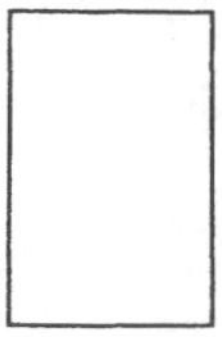	Intact rock	Brittle, elastic and generally isotropic behaviour	Triaxial testing of core specimens relatively simple and inexpensive and results are usually reliable	Behaviour of elastic isotropic rock is adequately understood for most practical applications
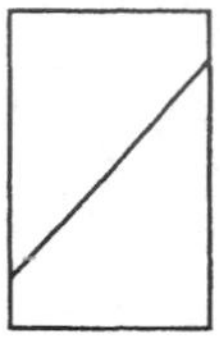	Intact rock with a single inclined discontinuity	Highly anisotropic, depending on shear strength and inclination of discontinuity	Triaxial tests difficult and expensive. Direct shear tests preferred. Careful interpretation of results required	Behaviour of discontinuities adequately understood for most practical applications
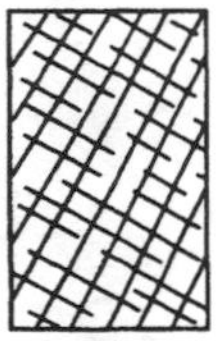	Massive rock with a few sets of discontinuities	Anisotropic, depend-ing on number, orientation and shear strength of discontinuities	Laboratory testing very difficult because of sample disturbance and equipment size limitations	Behaviour of complex block interaction in sparsely jointed rock masses poorly understood
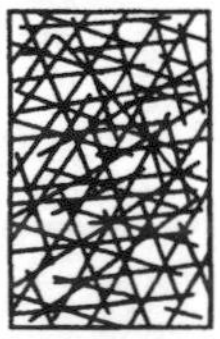	Heavily jointed rock masses	Reasonably isotropic, highly dilatant at low stress levels with particle breakage at high stress levels	Triaxial testing of representative samples extremely difficult because of sample disturbance	Behaviour of interlocking angular pieces poorly understood
	Compacted rock-fill or weakly cemented conglomerates	Reasonably isotropic, less dilatant and lower strength than in situ rock due to destruction of fabric	Triaxial testing simple but expensive due to large equipment required to accommodate samples	Behaviour reasonably well understood from soil mechanics studies on granular materials
	Loose waste rock or gravel	Poor compaction and grading allow particle movement resulting in mobility and low strength	Triaxial or direct shear testing simple but expensive due to large size of equipment	Behaviour of loosely compacted waste rock and gravel adequately understood for most applications

numerous alternatives lies in the adequacy of its predictions of observed rock fracture behaviour, and the convenience of its application to a range of typical engineering problems.

The Hoek-Brown failure criterion for intact rock may be expressed in the following form:

$$\sigma_1' = \sigma_3' + \sigma_c \left(m_i \frac{\sigma_3'}{\sigma_c} + 1 \right)^{1/2} \qquad (8.1)$$

where σ_1' is the major principal effective stress at failure
σ_3' is the minor principal effective stress at failure
σ_c is the uniaxial compressive strength of the intact rock
m_i is a material constant for the intact rock.

Whenever possible, the value of σ_c should be determined by laboratory testing on cores of approximately 50 mm diameter and 100 mm in length. In some cases, where the individual pieces of intact rock are too small to permit samples of this size to be tested, smaller diameter cores may be tested. Hoek and Brown (1980a) suggested that the equivalent uniaxial compressive strength of a 50 mm diameter core specimen can be estimated from:

$$\sigma_c = \frac{\sigma_{cd}}{\left(50/d\right)^{0.18}} \tag{8.2}$$

where σ_{cd} is the uniaxial strength measured on a sample of d mm in diameter.

The most reliable values of both the uniaxial compressive strength σ_c and the material constant m_i are obtained from the results of triaxial tests. For typical igneous and metamorphic rocks and for strong sedimentary rocks, such as sandstones, these laboratory tests are routine and there are many laboratories around the world which have excellent facilities for triaxial testing. In weak sedimentary rocks, such as shales and siltstones, preparation of specimens for triaxial testing can be very difficult because of the tendency of these materials to slake and de-laminate, when subjected to changes in moisture content. A solution which has been used on several major engineering projects is to carry out the triaxial tests in the field, usually in exploration adits or access tunnels, using a triaxial cell described by Franklin and Hoek (1970) and illustrated in Figure 8.1. This cell has a rubber sealing sleeve, which is designed to contain the pressurising fluid (usually oil), so that there is no need for drainage between tests. A diamond saw is used to trim the ends of the core sample and a capping compound is applied to produce parallel ends. A 50 ton capacity load frame provides a sufficiently high axial load for most of these weak rocks. Confining pressure is provided by a simple hand operated pump.

The specimen should be cored normal to significant discontinuities, such as bedding planes, and the tests should be carried out on specimens which have a moisture content as close to in situ conditions as possible. Although it is possible to obtain porous platens so that pore fluid pressures can be controlled, this control is not practical in field testing situations and a reasonable compromise is to keep loading rates low in order to avoid generation of dynamic pore pressures.

The triaxial test results can be processed using a program called ROCKDATA[1] developed by Shah (1992). This program is based upon the simplex reflection statistical technique which has been found to produce the most reliable interpretation of triaxial test data.

When time or budget constraints do not allow a triaxial testing programme to be carried out, the values of the constants σ_c and m_i

[1] Available from The Rock Engineering Group, 12 Selwood Avenue, Toronto, Ontario, Canada M4E 1B2, Fax 1 416 698 0908, Phone 1 416 698 8217.

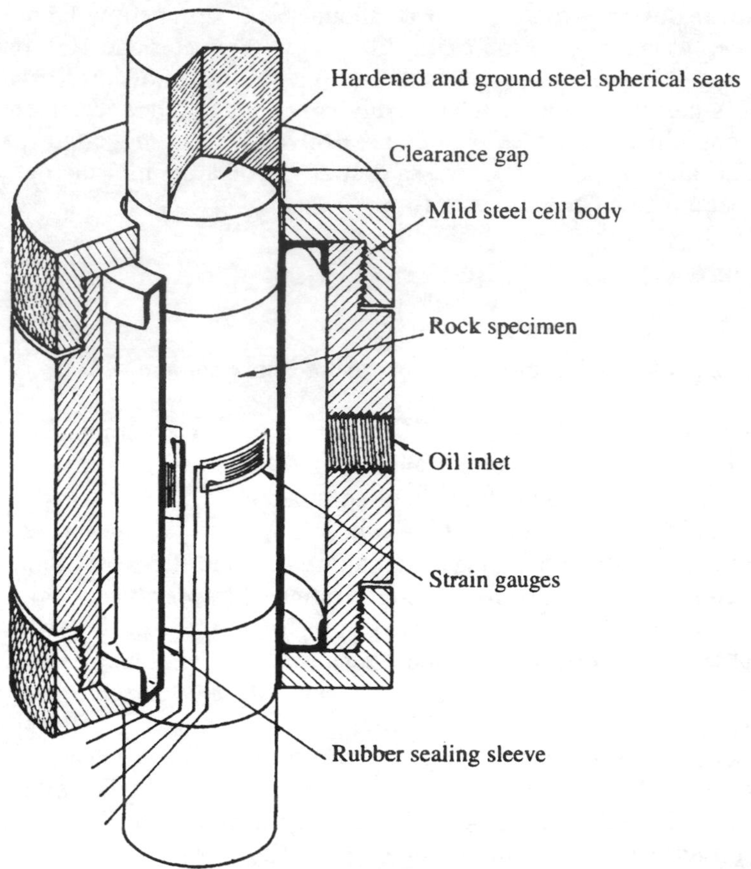

Figure 8.1: Simple triaxial cell used for testing rock cores in field laboratories. The rubber sealing sleeve is designed to retain the oil so that the cell does not need to be drained between tests. Cells are available to accommodate a variety of standard core sizes.

can be estimated from Tables 8.2 and 8.3. Table 8.3 is based upon analyses of published triaxial test results on intact rock (Hoek, 1983, Doruk, 1991 and Hoek et al., 1992).

A detailed discussion on the characteristics and limitations of the Hoek-Brown failure criterion, including the transition from brittle to ductile failure and the mechanics of anisotropic failure, has been given by Hoek (1983). These considerations are very important in the application of the failure criterion to the behaviour of intact rock. They may need to be considered when dealing with foliated rocks such as gneisses, which can exhibit strongly anisotropic behaviour, or with sedimentary rocks such as limestones and marbles, which may become ductile at low stress levels. However, in the context of this chapter, these detailed considerations are of secondary importance and will not be discussed further.

Table 8.2: Field estimates of uniaxial compressive strength.

Grade*	Term	Uniaxial comp. strength (MPa)	Point load index (MPa)	Field estimate of strength	Examples**
R6	Extremely strong	> 250	>10	Rock material only chipped under repeated hammer blows, rings when struck	Fresh basalt, chert, diabase, gneiss, granite, quartzite
R5	Very strong	100-250	4-10	Requires many blows of a geological hammer to break intact rock specimens	Amphibolite, sandstone, basalt, gabbro, gneiss, granodiorite, limestone, marble, rhyolite, tuff
R4	Strong	50-100	2-4	Hand held specimens broken by a single blow of geological hammer	Limestone, marble, phyllite, sandstone, schist, shale
R3	Medium strong	25-50	1-2	Firm blow with geological pick indents rock to 5 mm, knife just scrapes surface	Claystone, coal, concrete, schist, shale, siltstone
R2	Weak	5-25	***	Knife cuts material but too hard to shape into triaxial specimens	Chalk, rocksalt, potash
R1	Very weak	1-5	***	Material crumbles under firm blows of geological pick, can be shaped with knife	Highly weathered or altered rock
R0	Extremely weak	0.25-1	***	Indented by thumbnail	Clay gouge

* Grade according to ISRM (1981).

**All rock types exhibit a broad range of uniaxial compressive strengths which reflect the heterogeneity in composition and anisotropy in structure. Strong rocks are characterised by well interlocked crystal fabric and few voids.

***Rocks with a uniaxial compressive strength below 25 MPa are likely to yield highly ambiguous results under point load testing.

8.4 The strength of jointed rock masses

The original Hoek Brown criterion was published in 1980 and, based upon experience in using the criterion on a number of projects, an updated version was published in 1988 (Hoek and Brown, 1988) and a modified criterion was published in 1992 (Hoek et al., 1992).

The most general form of the Hoek-Brown criterion, which incorporates both the original and the modified form, is given by the equation

$$\sigma_1' = \sigma_3' + \sigma_c \left(m_b \frac{\sigma_3'}{\sigma_c} + s \right)^a \tag{8.3}$$

where m_b is the value of the constant m for the rock mass
s and a are constants which depend upon the characteristics of the rock mass
σ_c is the uniaxial compressive strength of the intact rock pieces and
σ_1' and σ_3' are the axial and confining effective principal stresses respectively.

The original criterion has been found to work well for most rocks of good to reasonable quality in which the rock mass strength is controlled by tightly interlocking angular rock pieces. The failure of such rock masses can be defined by setting $a = 0.5$ in Equation 8.3, giving

Table 8.3: Values of the constant m_i for intact rock, by rock group. Note that values in parenthesis are estimates.

Rock type	Class	Group	Texture			
			Course	Medium	Fine	Very fine
SEDIMENTARY	Clastic		Conglomerate (22)	Sandstone 19	Siltstone 9	Claystone 4
				← Greywacke (18) →		
	Non-Clastic	Organic		← Chalk 7 →		
				← Coal (8-21) →		
		Carbonate	Breccia (20)	Sparitic Limestone (10)	Micritic Limestone 8	
		Chemical		Gypstone 16	Anhydrire 13	
METAMORPHIC	Non Foliated		Marble 9	Hornfels (19)	Quartzite 24	
	Slightly foliated		Migmatite (30)	Amphibolite 31	Mylonites (6)	
	Foliated*		Gneiss 33	Schists (10)	Phyllites (10)	Slate 9
IGNEOUS	Light		Granite 33		Rhyolite (16)	Obsidian (19)
			Granodiorite (30)		Dacite (17)	
	Dark		Diorite (28)		Andesite 19	
			Gabbro 27	Dolerite (19)	Basalt (17)	
			Norite 22			
	Extrusive pyroclastic type		Agglomerate (20)	Breccia (18)	Tuff (15)	

*These values are for intact rock specimens tested normal to foliation. The value of m_i will be significantly different if failure occurs along a foliation plane (Hoek, 1983).

$$\sigma_1' = \sigma_3' + \sigma_c \left(m_b \frac{\sigma_3}{\sigma_c} + s \right)^{0.5} \qquad (8.4)$$

For poor quality rock masses in which the tight interlocking has been partially destroyed by shearing or weathering, the rock mass has no tensile strength or 'cohesion' and specimens will fall apart without confinement. For such rock masses the modified criterion is more appropriate and this is obtained by putting $s = 0$ in Equation 8.3 which gives:

$$\sigma_1' = \sigma_3' + \sigma_c \left(m_b \frac{\sigma_3}{\sigma_c} \right)^a \tag{8.5}$$

It is practically impossible to carry out triaxial or shear tests on rock masses at a scale which is appropriate for surface or underground excavations in mining or civil engineering. Numerous attempts have been made to overcome this problem by testing small scale models, made up from assemblages of blocks or elements of rock or of carefully designed model materials. While these model studies have provided a great deal of valuable information, they generally suffer from limitations arising from the assumptions and simplifications, which have to be made in order to permit construction of the models. Consequently, our ability to predict the strength of jointed rock masses on the basis of direct tests or of model studies is severely limited.

Equations 8.4 and 8.5 are of no practical value unless the values of the material constants m_b, s and a can be estimated in some way. Hoek and Brown (1988) suggested that these constants could be estimated from the 1976 version of Bieniawski's Rock Mass Rating (*RMR*), assuming completely dry conditions and a very favourable joint orientation. While this process is acceptable for rock masses with *RMR* values of more than about 25, it does not work for very poor rock masses since the minimum value which *RMR* can assume is 18. In order to overcome this limitation, a new index called the Geological Strength Index (*GSI*) is introduced. The value of *GSI* ranges from about 10, for extremely poor rock masses, to 100 for intact rock. The relationships between *GSI* and the rock mass classifications of Bieniawski and Barton, Lein and Lunde will be discussed in a later section of this chapter.

The relationships between m_b/m_i, s and a and the Geological Strength Index (*GSI*) are as follows:

For *GSI* > 25 (Undisturbed rock masses)

$$\frac{m_b}{m_i} = \exp\left(\frac{GSI - 100}{28} \right) \tag{8.6}$$

$$s = \exp\left(\frac{GSI - 100}{9} \right) \tag{8.7}$$

$$a = 0.5 \tag{8.8}$$

For *GSI* < 25 (Undisturbed rock masses)

$$s = 0 \tag{8.9}$$

$$a = 0.65 - \frac{GSI}{200} \tag{8.10}$$

Since many of the numerical models and limit equilibrium analyses used in rock mechanics are expressed in terms of the Mohr-Coulomb failure criterion, it is necessary to estimate an equivalent set of cohesion and friction parameters for given Hoek-Brown values. This can be done using a solution published by Balmer (1952) in which the normal and shear stresses are expressed in terms of the corresponding principal stresses as follows:

$$\sigma_n = \sigma_3 + \frac{\sigma_1 - \sigma_3}{\partial\sigma_1/\partial\sigma_3 + 1} \tag{8.11}$$

$$\tau = (\sigma_n - \sigma_3)\sqrt{\partial\sigma_1/\partial\sigma_3} \tag{8.12}$$

For the *GSI* > 25, when $a = 0.5$:

$$\frac{\partial\sigma_1}{\partial\sigma_3} = 1 + \frac{m_b\sigma_c}{2(\sigma_1 - \sigma_3)} \tag{8.13}$$

For *GSI* < 25, when $s = 0$:

$$\frac{\partial\sigma_1}{\partial\sigma_3} = 1 + a m_b^a \left(\frac{\sigma_3}{\sigma_c}\right)^{a-1} \tag{8.14}$$

Once a set of (σ_n, τ) values have been calculated from Equations 8.11 and 8.12, average cohesion c and friction angle ϕ values can be calculated by linear regression analysis, in which the best fitting straight line is calculated for the range of (σ_n, τ) pairs.

The uniaxial compressive strength of a rock mass defined by a cohesive strength c and a friction angle ϕ *is* given by:

$$\sigma_{cm} = \frac{2c.\cos\phi}{1 - \sin\phi} \tag{8.15}$$

A simple spreadsheet for carrying out the full range of calculations presented above is given in Figure 8.2.

8.5 Use of rock mass classifications for estimating GSI

In searching for a solution to the problem of estimating the strength of jointed rock masses and to provide a basis for the design of underground excavations in rock, Hoek and Brown (1980a, 1980b) felt that some attempt had to be made to link the constants m and s of their criterion to measurements or observations which could be carried out by any competent geologist in the field. Recognising that the characteristics of the rock mass which control its strength and deformation behaviour are similar to the characteristics, which had been adopted by Bieniawski (1973) and by Barton et al. (1974) for their rock mass classifications, Hoek and Brown proposed that these classifications could be used for estimating the material constants m and s.

In preparing the present book it became obvious that there was a need to consolidate these various versions of the criterion into a single simplified and generalised criterion to cover all of the rock types normally encountered in underground engineering.

The rock mass classifications by Bieniawski (1974) and Barton et al. (1974) were developed for the estimation of tunnel support. They were adopted by Hoek and Brown (1980) for estimating m and s values because they were already available and well established in 1980, and because there appeared to be no justification for proposing yet another classification system. However, there is a potential problem in using these existing rock mass classification systems as a basis for estimating the strength of a rock mass.

Consider a tunnel in a highly jointed rock mass subjected to an in situ stress field such that failure can occur in the rock surrounding the tunnel. When using the Tunnelling Quality Index Q proposed by Barton et al. (1974) for estimating the support required for the tunnel, the in situ stress field is allowed for by means of a Stress Reduction Factor. This factor can have a significant influence upon the level of support recommended on the basis of the calculated value of Q. An alternative approach to support design is to estimate the strength of the rock mass by means of the Hoek-Brown failure criterion. This strength is then applied to the results of an analysis of the stress distribution around the tunnel, in order to estimate the extent of zones of overstressed rock requiring support. If the Barton et al. classification has been used to estimate the values of m and s, and if the Stress Reduction Factor has been used in calculating the value of Q, it is clear that the influence of the in situ stress level will be accounted for twice in the analysis.

ESTIMATE OF HOEK-BROWN AND MOHR-COULOMB PARAMETERS

Input : GSI = 62 sigci = 100 mi = 24

Output:		sig3	sig1	ds1ds3	sign	tau	signtau	signsq
mb/mi =	0.26	0.10	14.48	22.47	0.71	2.91	2.07	0.51
mb =	6.18	0.20	16.55	19.89	0.98	3.49	3.41	0.96
s =	0.015	0.39	20.09	16.68	1.50	4.55	6.85	2.26
a =	0.5	0.78	25.87	13.31	2.53	6.39	16.20	6.42
E =	19953	1.56	34.91	10.26	4.52	9.48	42.90	20.46
phi =	48	3.13	48.70	7.78	8.32	14.48	120.44	69.18
coh =	3.4	6.25	69.56	5.88	15.45	22.31	344.80	238.78
sigcm =	18.0	12.5	101.20	4.48	28.68	34.26	982.51	822.60
				Sums =	62.70	97.88	1519.17	1161.16

Cell formulae:

```
   mb/mi = EXP((GSI-100)/28)
      mb = mi*EXP((GSI-100)/28)
       s = IF(GSI>25 THEN EXP((GSI-100)/9) ELSE 0)
       a = IF(GSI>25 THEN 0.5 ELSE (0.65-GSI/200))
       E = 1000*10^((GSI-10)/40)
    sig3 = sigci/2^n where n starts at 10 and decreases by 1 for each subsequent cell
    sig1 = sig3+sigci*(((mb*sig3)/sigci) + s)^a
  ds1ds3 = IF(GSI>25 THEN 1+(mb*sigci)/(2*(sig1-sig3)) ELSE 1+(a*mb^a)*(sig3/sigci)^(a-1))
    sign = sig3+(sig1-sig3)/(1+ds1ds3)
     tau = (sign-sig3)*SQRT(ds1ds3)
 signtau = sign*tau                  signsq = sign^2
     phi = (ATAN((sum(signtau)-(sum(sign)*sum(tau))/8)/(sum(signsq)-((sum(sign))^2)/8)))*180/PI()
     coh = (sum(tau)/8) - (sum(sign)/8)*TAN(phi*PI()/180)
   sigcm = (2*coh*COS(phi*PI()/180))/(1-SIN(phi*PI()/180))
```

Figure 8.2: Spreadsheet for the calculation of Hoek-Brown and Mohr-Coulomb parameters.

Table 8.4: Estimation of constants m_b /m_i, s, a, deformation modulus E and the Poisson s ratio ν for the Generalised Hoek-Brown failure criterion based upon rock mass structure and discontinuity surface conditions. Note that the values given in this table are for an *undisturbed* rock mass.

GENERALISED HOEK-BROWN CRITERION

$$\sigma_1' = \sigma_3' + \sigma_c \left(m_b \frac{\sigma_3'}{\sigma_c} + s \right)^a$$

σ_1' = major principal effective stress at failure

σ_3' = minor principal effective stress at failure

σ_c = uniaxial compressive strength of *intact* pieces of rock

m_b , s and a are constants which depend on the composition, structure and surface conditions of the rock mass

STRUCTURE	SURFACE CONDITION	VERY GOOD Very rough, unweathered surfaces	GOOD Rough, slightly weathered, iron stained surfaces	FAIR Smooth, moderately weathered or altered surfaces	POOR Slickensided, highly weathered surfaces with compact coatings or fillings containing angular rock fragments	VERY POOR Slickensided, highly weathered surfaces with soft clay coatings or fillings
BLOCKY -very well interlocked undisturbed rock mass consisting of cubical blocks formed by three orthogonal discontinuity sets	m_b/m_i	0.60	0.40	0.26	0.16	0.08
	s	0.190	0.062	0.015	0.003	0.0004
	a	0.5	0.5	0.5	0.5	0.5
	E_m	75,000	40,000	20,000	9,000	3,000
	ν	0.2	0.2	0.25	0.25	0.25
	GSI	85	75	62	48	34
VERY BLOCKY-interlocked, partially disturbed rock mass with multifaceted angular blocks formed by four or more discontinuity sets	m_b/m_i	0.40	0.29	0.16	0.11	0.07
	s	0.062	0.021	0.003	0.001	0
	a	0.5	0.5	0.5	0.5	0.53
	E_m	40,000	24,000	9,000	5,000	2,500
	ν	0.2	0.25	0.25	0.25	0.3
	GSI	75	65	48	38	25
BLOCKY/SEAMY-folded and faulted with many intersecting discontinuities forming angular blocks	m_b/m_i	0.24	0.17	0.12	0.08	0.06
	s	0.012	0.004	0.001	0	0
	a	0.5	0.5	0.5	0.5	0.55
	E_m	18,000	10,000	6,000	3,000	2,000
	ν	0.25	0.25	0.25	0.3	0.3
	GSI	60	50	40	30	20
CRUSHED-poorly interlocked, heavily broken rock mass with a mixture of angular and rounded blocks	m_b/m_i	0.17	0.12	0.08	0.06	0.04
	s	0.004	0.001	0	0	0
	a	0.5	0.5	0.5	0.55	0.60
	E_m	10,000	6,000	3,000	2,000	1,000
	ν	0.25	0.25	0.3	0.3	0.3
	GSI	50	40	30	20	10

Note 1: The in situ deformation modulus E_m is calculated from Equation 4.7 (page 47, Chapter 4). Units of E_m are MPa.

Similar considerations apply to the Joint Water Reduction Factor in Barton et al.'s classification and to the Ground Water term and the Rating Adjustment for Joint Orientations in Bieniawski's *RMR* classification. In all cases there is a potential for double counting, if these factors are not treated with care when using these classifications as a basis for estimating the strength of rock masses.

In order to minimise potential problems of the type described above, the following guidelines are offered for the selection of parameters when using rock mass classifications as a basis for estimating m and s values for the Hoek-Brown failure criterion.

Bieniawski's 1976 RMR classification

Bieniawski has made several changes to the ratings used in his classification (Bieniawski, 1973, 1974, 1976, 1979, 1989) and the significance of these changes is best appreciated by considering the following typical example:

A slightly weathered granite has an average Point-load strength index value of 7 MPa, an average *RQD* value of 70%, and slightly rough joints with a separation of < 1 mm, are spaced at 300 mm. The *RMR* values for this rock mass, calculated using tables published by Bieniawski in the years indicated, are as follows:

Item	Value	1973	1974	1976	1979	1989
Point load index	7 MPa	5	5	12	12	12
RQD	70%	14	14	13	13	13
Spacing of discontinuities	300 mm	20	20	20	10	10
Condition of discontinuities	Described	12	10	20	20	25
Groundwater	Dry	10	10	10	15	15
Joint orientation adjustment	Very favourable	15	15	0	0	0
	RMR	76	74	75	70	75

The differences in these values demonstrate that it is essential that the correct ratings be used. The 1976 paper by Bieniawski is the basic reference for this work. For the convenience of the reader, the relevant parts of Bieniawski's 1976 Geomechanics Classification table are reproduced in Table 8.3.

In using Bieniawski's 1976 Rock Mass Rating to estimate the value of *GSI*, Table 8.3 should be used to calculate the ratings for the first four terms. The rock mass should be assumed to be completely dry and a rating of 10 assigned to the Groundwater value. Very favourable joint orientations should be assumed and the Adjustment for Joint Orientation value set to zero. The final rating, called RMR_{76}', can then be used to estimate the value of *GSI*:

For $RMR_{76}' > 18$

$$GSI = RMR_{76}' \tag{8.16}$$

For $RMR_{76}' < 18$ Bieniawski's 1976 classification cannot be used to estimate *GSI* and Barton, Lein and Lunde's Q' value should be used instead.

Table 8.5: Part of Bieniawski's 1976 table defining the Geomechanics Classification or Rock Mass Rating (*RMR*) for jointed rock masses.

	PARAMETER		**RANGE OF VALUES**						
1	Strength of intact rock material	Point-load strength index	>8 MPa	4-8 MPa	2-4 MPa	1-2 MPa	For this low rangeuniaxial compressive test is preferred		
		Uniaxial compressive strength	>200 MPa	100-200 MPa	50-100 MPa	25-50 MPa	10-25 MPa	3-10 MPa	1-3 MPa
	Rating		15	12	7	4	2	1	0
2	Drill core quality *RQD*		90%- 100%	75%- 90%	50%-75%	25%-50%	< 25%		
	Rating		20	17	13	8	3		
3	Spacing of joints		> 3 m	1-3 m	0.3-1 m	50-300 mm	< 50 mm		
	Rating		30	25	20	10	5		
4	Condition of joints		Very rough surfaces Not continuous No separation Hard joint wall contact	Slightly rough surfaces Separation < 1 mm Hard joint wall contact	Slightly rough surfaces Separation < 1 mm Soft joint wall contact	Slickensided surfaces or Gouge < 5 mm thick or Joints open 1-5 mm Continuous joints	Soft gouge >5 mm thick or Joints open > 5 mm Continuous joints		
	Rating		25	20	12	6	0		

Bieniawski's 1989 RMR classification

Bieniawski's 1989 classification, given in Table 4.4 on page 35, can be used to estimate the value of *GSI* in a similar manner to that described above for the 1976 version. In this case a value of 15 is assigned to the Groundwater rating and the Adjustment for Joint Orientation is again set to zero. Note that the minimum value which can be obtained for the 1989 classification is 23 and that, in general, it gives a slightly higher value than the 1976 classification. The final rating, called RMR_{89}', can be used to estimate the value of *GSI*:

For $RMR_{89}' > 23$

$$GSI = RMR_{89}' - 5 \tag{8.17}$$

For $RMR_{89}' < 23$ Bieniawski's 1976 classification cannot be used to estimate *GSI* and Barton, Lein and Lunde's *Q'* value should be used instead.

Modified Barton, Lien and Lunde's Q' classification

In using this classification to estimate *GSI*, the Rock Quality Designation (*RQD*), joint set number (J_n), joint roughness number (J_r) and joint alteration number (J_a) should be used exactly as defined in the tables published by Barton et al. (1974) and given in Table 4.6 on pages 41 to 43.

For the joint water reduction factor (J_w) and the stress reduction factor (*SRF*), use a value of 1 for both of these parameters, equivalent to a dry rock mass subjected to medium stress conditions. The influence of both water pressure and stress should be included in the analysis of stresses acting on the rock mass for which failure is defined in terms of the Hoek-Brown failure criterion.

Hence, for substitution into Equation 8.7, the modified Tunnelling Quality Index (*Q'*) is calculated from:

$$Q' = \frac{RQD}{J_n} \times \frac{J_r}{J_a} \tag{8.18}$$

This value of Q' can be used to estimate the value of *GSI* from:

$$GSI = 9 \text{Log}_e Q' + 44 \tag{8.19}$$

Note that the minimum value for Q' is 0.0208 which gives a *GSI* value of approximately 9 for a thick, clay-filled fault or shear zone.

8.6 When to use the Hoek-Brown failure criterion

The rock mass conditions under which the Hoek-Brown failure criterion can be applied are summarised in Figure 8.3.

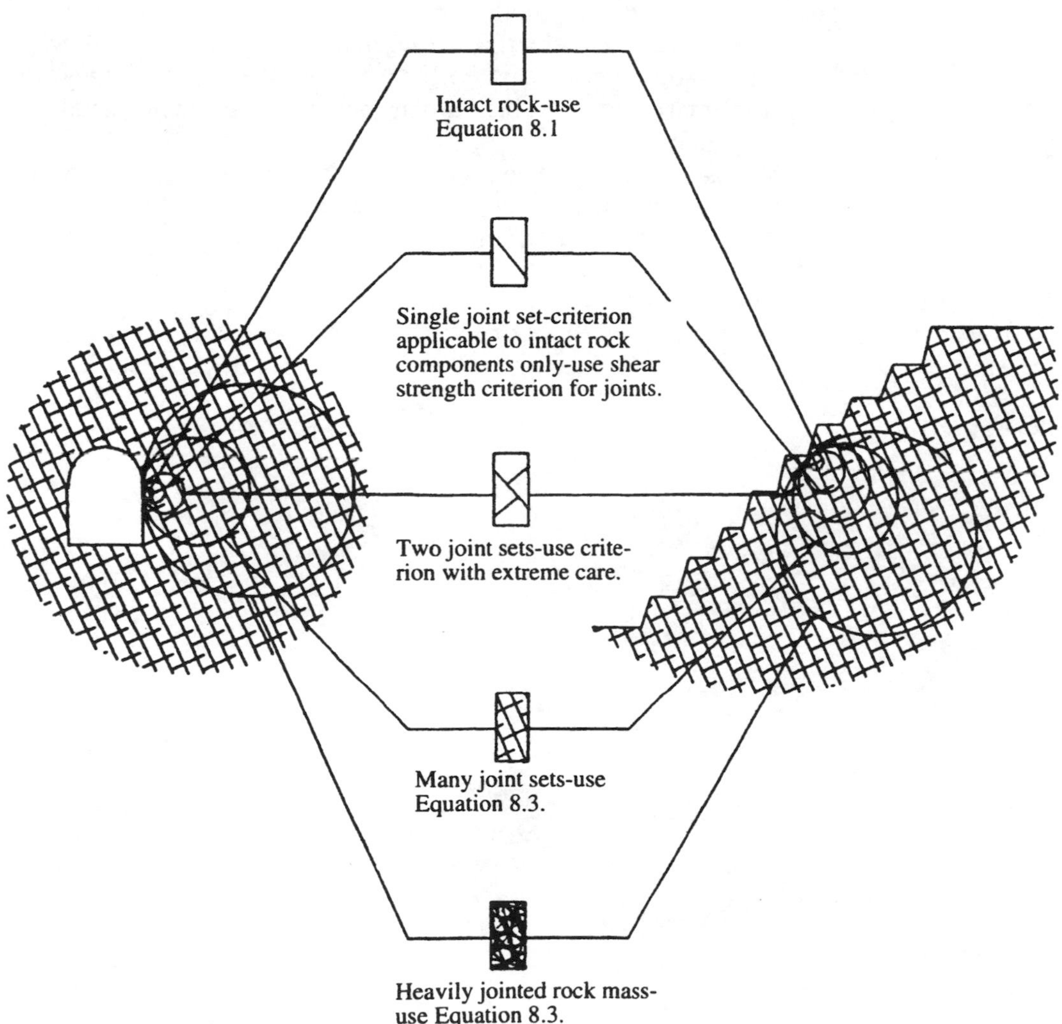

Figure 8.3: Rock mass conditions under which the Hoek-Brown failure criterion can be applied.

The Hoek-Brown failure criterion is only applicable to intact rock or to heavily jointed rock masses which can be considered homogeneous and isotropic. In other words the properties of these materials are the same in all directions.

The criterion should not be applied to highly schistose rocks such as slates or to rock masses in which the properties are controlled by a single set of discontinuities such as bedding planes. In cases where such rock masses are being analysed, the Hoek-Brown failure criterion applies to the intact rock components only. The strength of the discontinuities should be analysed in terms of the shear strength criteria discussed in Chapter 5.

When two joint sets occur in a rock mass, the Hoek-Brown criterion can be used with extreme care, provided that neither of the joint sets has a dominant influence on the behaviour of the rock mass. For example, if one of the joint sets is clay coated and is obviously very much weaker than the other set, the Hoek-Brown criterion should not be used except for the intact rock components. On the other hand, when both joint sets are fresh, rough and unweathered and when their orientation is such that no local wedge failures are likely, the upper left hand box in Table 8.4 can be used to estimate the Hoek-Brown parameters.

For more heavily jointed rock masses in which many joints occur, the Hoek-Brown criterion can be applied and Table 8.4 can be used to estimate the strength parameters.

9 Support design for overstressed rock

9.1 Introduction

The failure of a rock mass around an underground opening depends upon the in situ stress level and upon the characteristics of the rock mass. Figure 9.1 gives a simplified description of the various types of failure which are commonly observed underground. The stability of structurally controlled failures in jointed rock masses and the design of support systems for this type of failure were dealt with in Chapter 6. In this chapter, the question of failure and the design of support for highly stressed rock masses will be discussed.

The right hand column of Figure 9.1 shows that failure around openings in highly stressed rock masses progresses from brittle spalling and slabbing, in the case of massive rocks with few joints, to a more ductile type of failure for heavily jointed rock masses. In the latter case, the presence of many intersecting discontinuities provides considerable freedom for individual rock pieces to slide or to rotate within the rock mass. The presence of clay gouge or of slickensided surfaces further weakens the rock mass and contributes to the ductile or 'plastic' failure of such rock masses. In the intermediate case, structure and intact rock failure combine to create a complex series of failure mechanisms. In situations with distinctly anisotropic strength, such as thinly bedded, folded or laminated rock, brittle failure processes such as buckling may occur.

In discussing the question of support design for overstressed rock, it is instructive to start with the lower right hand box in Figure 9.1 to consider how a heavily jointed rock mass fails, and how installed support reacts to the displacements induced by this failure.

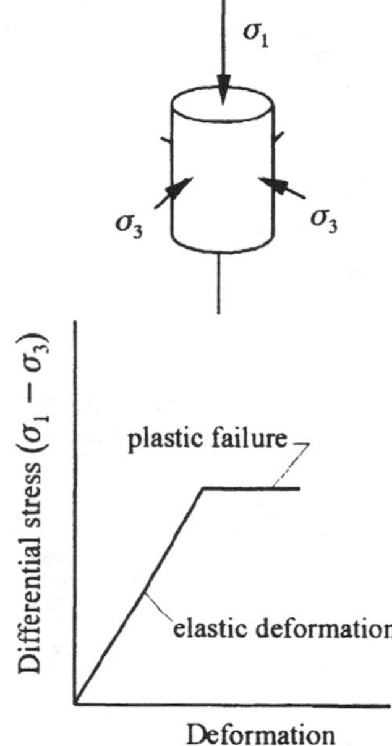

Stress-deformation curve for a constant confining pressure

9.2 Support interaction analysis

In order to present the concepts of rock support interaction in a form which can be readily understood, a very simple analytical model will be utilised. This model involves a circular tunnel subjected to a hydrostatic stress field in which the horizontal and vertical stresses are equal. The surrounding rock mass is assumed to behave as an elastic-perfectly plastic material as illustrated in the margin sketch. Failure, involving slip along intersecting discontinuities in a heavily jointed rock mass, is assumed to occur with zero plastic volume change (Duncan Fama, 1993). Support is modelled as an equivalent internal pressure, hence, the reinforcement provided by grouted rockbolts or cables cannot be taken into account in this simple model.

	Low stress levels	High stress levels
Massive rock	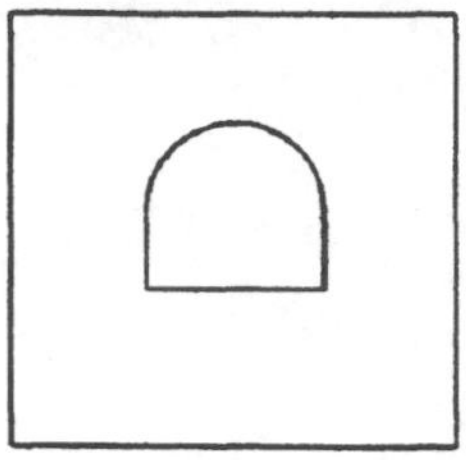 Massive rock subjected to low in situ stress levels. Linear elastic response with little or no rock failure.	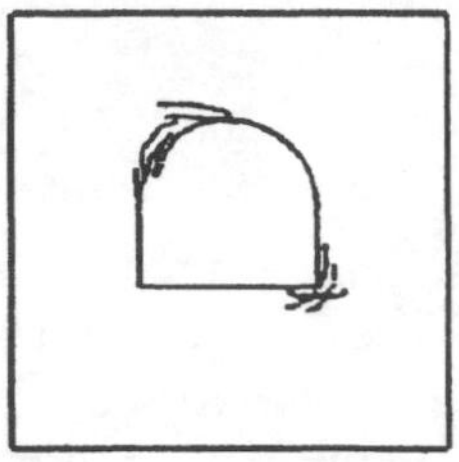 Massive rock subjected to high in situ stress levels. Spalling, slabbing and crushing initiates at high stress concentration points on the boundary and propagates into the surrounding rock mass.
Jointed rock	 Massive rock, with relatively few discontinuities, subjected to low in situ stress conditions. Blocks or wedges, released by intersecting discontinuities, fall or slide due to gravity loading.	 Massive rock, with relatively few discontinuities, subjected to high in situ stress conditions. Failure occurs as a result of sliding on discontinuity surfaces and also by crushing and splitting of rock blocks.
Heavily jointed rock	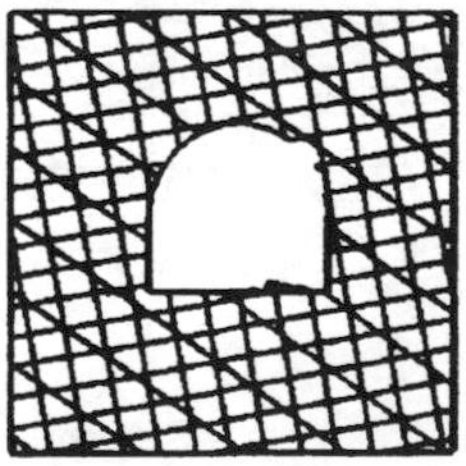 Heavily jointed rock subjected to low in situ stress conditions. The opening surface fails as a result of unravelling of small interlocking blocks and wedges. Failure can propagate a long way into the rock mass if it is not controlled.	 Heavily jointed rock subjected to high in situ stress conditions. The rock mass surrounding the opening fails by sliding on discontinuities and crushing of rock pieces. Floor heave and sidewall closure are typical results of this type of failure.

Figure 9.1: Types of failure which occur in different rock masses under low and high in situ stress levels.

9.2.1 *Definition of failure criterion*

It is assumed that the onset of plastic failure, for different values of the confining stress σ_3 is defined by the Mohr-Coulomb criterion which may be expressed as:

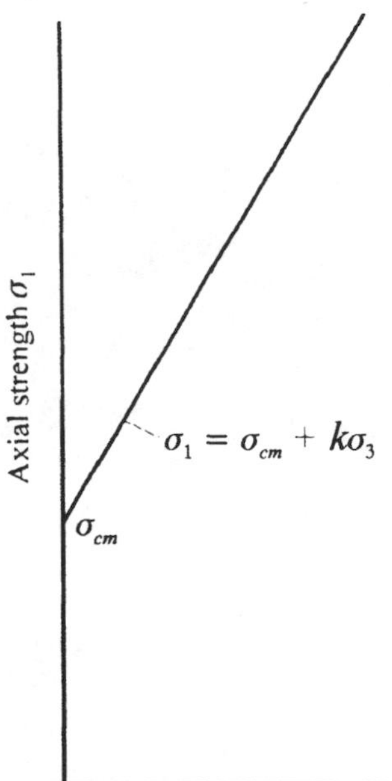

$$\sigma_1 = \sigma_{cm} + k\sigma_3 \tag{9.1}$$

The uniaxial compressive strength of the rock mass σ_{cm} is defined by:

$$\sigma_{cm} = \frac{2c\,\cos\phi}{(1-\sin\phi)} \tag{9.2}$$

and the slope k of σ_1 the versus σ_3 line as:

$$k = \frac{(1+\sin\phi)}{(1-\sin\phi)} \tag{9.3}$$

where σ_1 is the axial stress at which failure occurs
σ_3 is the confining stress
c is the cohesive strength and
ϕ is the angle of friction of the rock mass

In order to estimate the cohesive strength c and the friction angle ϕ for an actual rock mass, the procedure outlined in Section 8.4 of the previous chapter can be utilised. Having estimated the parameters for the Hoek-Brown failure criterion as described in that section, values for c and ϕ can be calculated by means of the spreadsheet given in Figure 8.2.

9.2.2 *Analysis of tunnel behaviour*

Assume that a circular tunnel of radius r_o subjected to hydrostatic stresses and a uniform internal support pressure p_i as illustrated in the margin sketch. Failure of the rock mass surrounding the tunnel occurs, when the internal pressure provided by the tunnel lining is less than a critical support pressure p_{cr}, which is defined by:

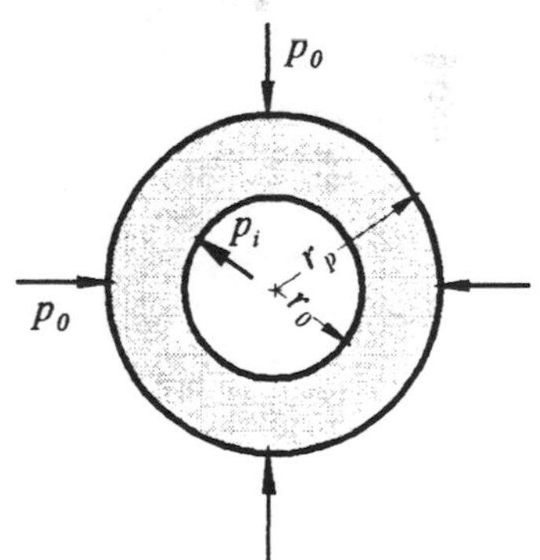

$$p_{cr} = \frac{2p_o - \sigma_{cm}}{1+k} \tag{9.4}$$

If the internal support pressure p_i is greater than the critical support pressure p_{cr}, no failure occurs and the behaviour of the rock mass surrounding the tunnel is elastic. The inward radial elastic displacement of the tunnel wall is given by:

$$u_{ie} = \frac{r_o(1+\nu)}{E}(p_o - p_i) \tag{9.5}$$

where E is the Young's modulus or deformation modulus and
ν is the Poisson's ratio.

When the internal support pressure p_i is less than the critical support pressure p_{cr}, failure occurs and the radius r_p of the plastic zone around the tunnel is given by:

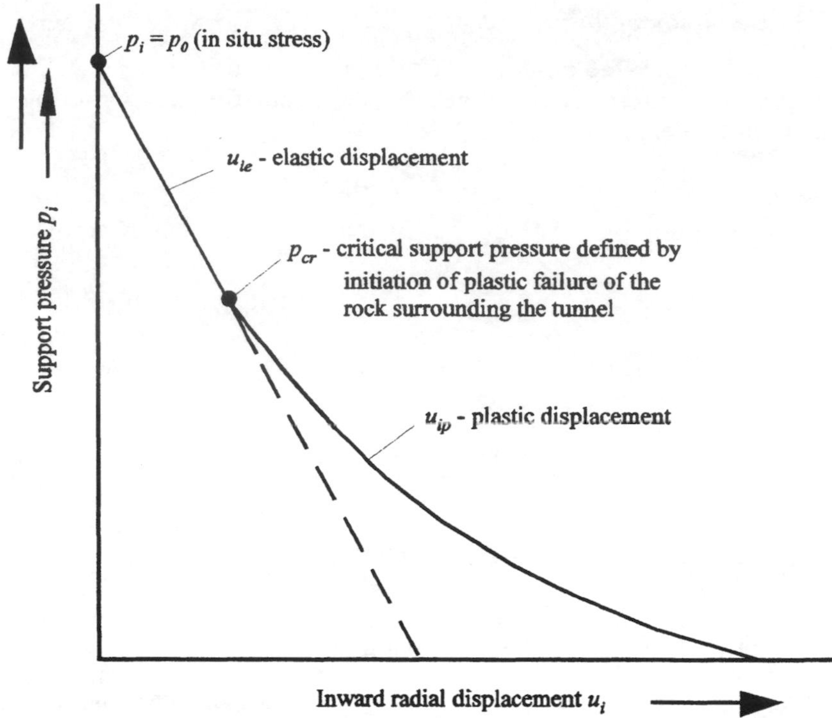

Figure 9.2: Graphical representation of relationships between support pressure and radial displacement of tunnel walls defined by Equations 9.5 and 9.7.

$$r_p = r_o \left[\frac{2(p_o(k-1)+\sigma_{cm})}{(1+k)((k-1)p_i + \sigma_{cm})} \right]^{\frac{1}{(k-1)}} \tag{9.6}$$

The total inward radial displacement of the walls of the tunnel is given by:

$$u_{ip} = \frac{r_o(1+\nu)}{E} \left[2(1-\nu)(p_o - p_{cr}) \left(\frac{r_p}{r_o} \right)^2 - (1-2\nu)(p_o - p_i) \right] \tag{9.7}$$

A typical plot of the displacements predicted by Equations 9.5 and 9.7 is given in Figure 9.2. This plot shows zero displacement when the support pressure equals the hydrostatic stress ($p_i = p_o$), elastic displacement for $p_o > p_i > p_{cr}$, plastic displacement for $p_i < p_{cr}$ and a maximum displacement when the support pressure equals zero.

9.2.3 *Deformation of an unsupported tunnel*

In order to understand how the support pressure operates, it is useful to start with an examination of Figure 9.3 which shows the response of the rock mass surrounding an advancing tunnel.

Consider the response of a measuring point installed well ahead of the advancing tunnel. Assume that no rockbolts, shotcrete lining or steel sets are installed and that the only support provided is by the rock ahead of the advancing face. Measurable displacement in the rock mass begins at a distance of about one half a tunnel diameter ahead of the face. The displacement increases gradually and, when

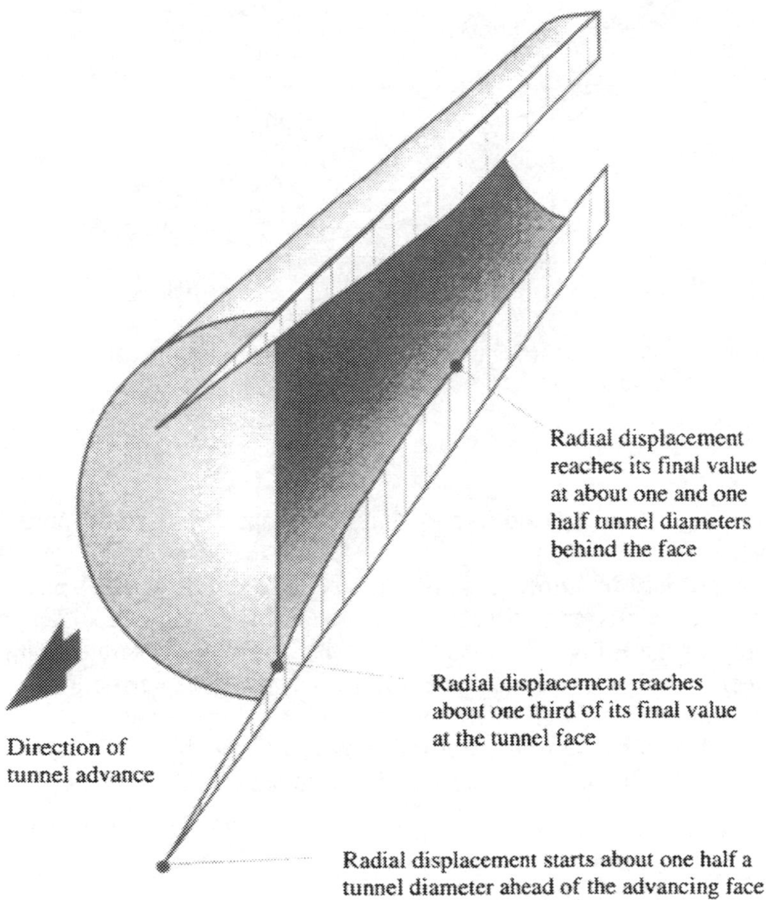
Radial displacement reaches its final value at about one and one half tunnel diameters behind the face
Radial displacement reaches about one third of its final value at the tunnel face
Direction of tunnel advance
Radial displacement starts about one half a tunnel diameter ahead of the advancing face

Figure 9.3: Pattern of radial deformation in the roof and floor of an advancing tunnel.

the tunnel face is coincident with the measuring point, the radial displacement is about one third of the maximum value. The displacement reaches a maximum when the face has progressed about one and one half tunnel diameters beyond the measuring point and the support provided by the face is no longer effective.

When the rock mass is strong enough to resist failure, i.e. when $\sigma_{cm} > 2p_o$ for $p_i = 0$ (from Equation 9.4), the displacements are elastic and follow the dashed line shown in Figure 9.2. When failure takes place, the displacements are plastic and follow the solid curve indicated in Figure 9.2.

Note that plastic failure of the rock mass surrounding the tunnel does not necessarily mean that the tunnel collapses. The failed material still has considerable strength and, provided that the thickness of the plastic zone is small compared with the tunnel radius, the only evidence of failure may be a few fresh cracks and a minor amount of ravelling or spalling. On the other hand, when a large plastic zone is formed and when large inward displacements of the tunnel wall occur, the loosening of the failed rock mass will lead to severe

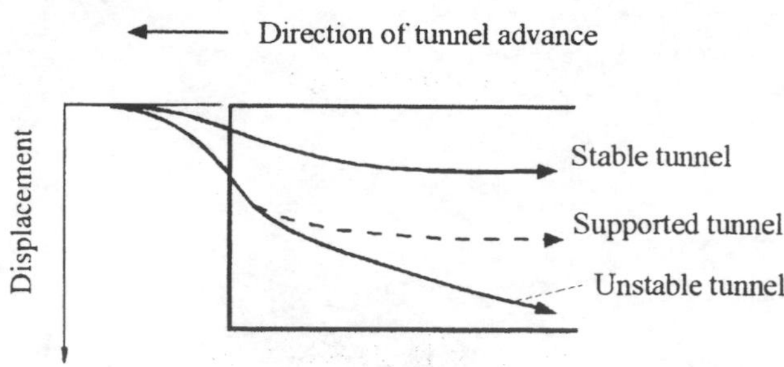

Figure 9.4: Displacement curves for the roof of a tunnel for different stability conditions in the surrounding rock mass.

spalling and ravelling and to an eventual collapse of an unsupported tunnel.

The primary function of support is to control the inward displacement of the walls and to prevent the loosening, which can lead to collapse of the tunnel. The installation of rockbolts, shotcrete lining or steel sets cannot prevent the failure of the rock surrounding a tunnel subjected to significant overstressing; but these support types do play a major role in controlling tunnel deformation. A graphical summary of this concept is presented in Figure 9.4.

9.2.4 *Deformation characteristics of support*

As illustrated in Figures 9.3 and 9.4, a certain amount of deformation occurs ahead of the advancing face of the tunnel. At the face itself, approximately one third of the total deformation has already occurred and this deformation cannot be recovered. In addition, there is almost always a stage of the excavation cycle in which there is a gap between the face and the closest installed support element. Hence, further deformation occurs before the support becomes effective. This total initial displacement will be called u_{so} and it is shown in Figure 9.5.

Once the support has been installed and it is in full and effective contact with the rock, the support starts to deform elastically as shown in Figure 9.5. The maximum elastic displacement which can be accommodated by the support system is u_{sm} and the maximum support pressure p_{sm} is defined by yield of the support system.

Depending upon the characteristics of the support system, the rock mass surrounding the tunnel and the in situ stress level, the support system will deform elastically in response to the closure of the tunnel, as the face advances away from the point under consideration. Equilibrium is achieved, if the support reaction curve intersects the rock mass displacement curve before either of these curves have progressed too far. If the support is installed too late (i.e. u_{so} is large in Figure 9.5), the rock mass may already have deformed to the extent that loosening of the failed material is irreversible. On the other hand, if the capacity of the support is inadequate (i.e. p_{sm} is low in Figure 9.5), then yield of the support may occur before the

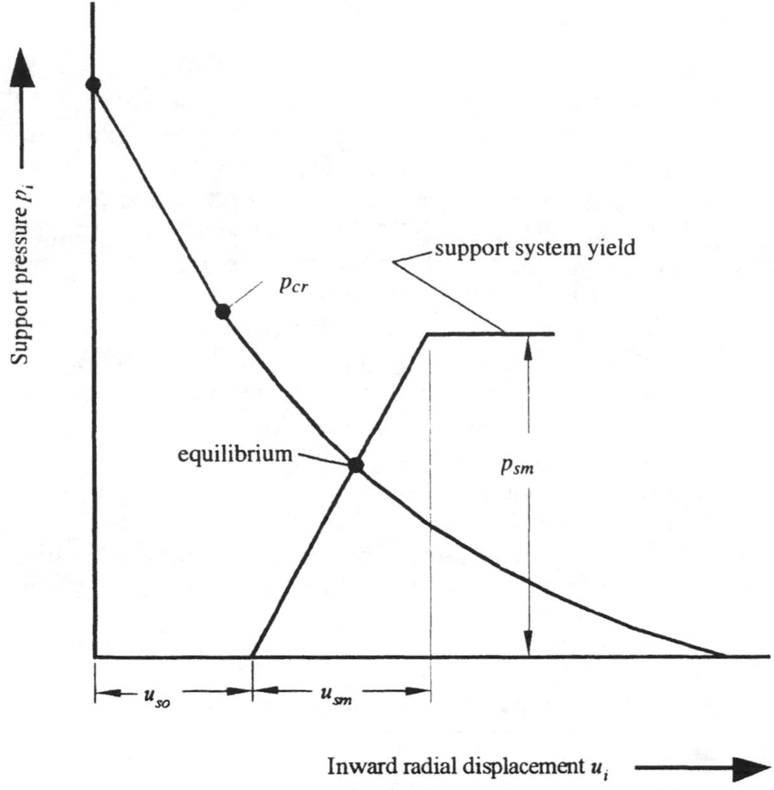

Figure 9.5: Response of support system to tunnel wall displacement resulting in establishment of equilibrium.

rock mass deformation curve is intersected. In either of these cases the support system will be ineffective, since the equilibrium condition, illustrated in Figure 9.5, will not have been achieved.

Because a number of factors are involved in defining the curves illustrated in Figure 9.5, it is very difficult to give general guidelines on the choice of support for every situation, even for this very simple case of a circular tunnel in a hydrostatic stress field. Some readers may argue that the analysis which has been presented is too simple to give meaningful results and that further discussion on this topic is not justified. However, the authors suggest that a great deal can be learned by carrying out parametric studies in which different combinations of in situ stress levels, rock mass strengths and support characteristics are evaluated. These parametric studies are most conveniently carried out by means of a spreadsheet program such as that presented in Figure 9.6.

Before discussing the operation of this program and the significance of the results which it produces, it is necessary to consider the question of the capacity of different support systems.

9.2.5 *Estimates of support capacity*

Hoek and Brown (1980a) and Brady and Brown (1985) have published equations which can be used to calculate the capacity of mechanically anchored rockbolts, shotcrete or concrete linings or steel sets for a circular tunnel. No useful purpose would be served by reproducing these equations here but they have been used to estimate the values listed in Table 9.1. This table gives maximum support pressures (p_{sm}) and maximum elastic displacements (u_{sm}) for different support systems installed in circular tunnels of different diameters.

Note that, in all cases, the supports are assumed to act over the entire surface of the tunnel walls. In other words, the shotcrete and concrete linings are closed rings; the steel sets are complete circles; and the mechanically anchored rockbolts are installed in a regular pattern which completely surrounds the tunnel.

Because this model assumes perfect symmetry under hydrostatic loading of circular tunnels, no bending moments are induced in the support. In reality, there will always be some asymmetric loading, particularly for steel sets and shotcrete placed on rough rock sur-

Table 9.1: Approximate support characteristics for different support systems installed in circular tunnels of various diameters.

Support type	Tunnel diameter - m	4	6	8	10	12
Very light rockbolts [1] 16mm dia.	Maximum pressure - MPa	0.25	0.11	0.06	0.04	0.03
Pullout load =0.11 MN	Max. elastic displacement - mm	10	12	13	14	15
Light rockbolts [1] 19mm diameter	Maximum pressure - MPa	0.40	0.18	0.10	0.06	0.04
Pullout load =0.18 MN	Max. elastic displacement - mm	12	14	15	17	18
Medium rockbolts [1] 25mm diameter	Maximum pressure - MPa	0.60	0.27	0.15	0.10	0.07
Pullout load = 0.27 MN	Max. elastic displacement - mm	15	16	17	19	20
Heavy rockbolts [1] 34mm diameter	Maximum pressure - MPa	0.77	0.34	0.19	0.12	0.09
Pullout load =0.35 MN	Max. elastic displacement - mm	19	21	22	23	24
One day old shotcrete 50mm [2]	Maximum pressure - MPa	0.35	0.23	0.17	0.14	0.12
UCS = 14 MPa, Ec= 8500 MPa	Max. elastic displacement - mm	3	5	6	8	10
28 day old shotcrete 50mm [2]	Maximum pressure - MPa	0.86	0.58	0.43	0.35	0.29
UCS = 35 MPa, Ec= 21000 MPa	Max. elastic displacement - mm	3	5	6	8	9
28 day old concrete 300mm	Maximum pressure - MPa	4.86	3.33	2.53	2.04	1.71
UCS = 35 MPa, Ec= 21000 MPa	Max. elastic displacement - mm	3	4	6	7	9
Light steel sets 6I12 [3]	Maximum pressure - MPa	0.33	0.18	0.12	0.08	0.06
Spaced at 1.5m, well blocked	Max. elastic displacement - mm	7	7	8	8	9
Medium steel sets 8I23 [4]	Maximum pressure - MPa	[6]	0.37	0.25	0.17	0.13
Spaced at 1.5m, well blocked	Max. elastic displacement - mm		8	9	10	10
Heavy steel sets 12W65 [5]	Maximum pressure - MPa	[6]	[6]	0.89	0.66	0.51
Spaced at 1.5m, well blocked	Max. elastic displacement - mm			9	11	12

Notes: [1]Rockbolts are mechanically anchored and ungrouted. Bolt length is assumed to be equal to 1/3 of the tunnel diameter and bolt spacing is one half bolt length. [2]Values apply to a completely closed shotcrete ring. For a shotcrete lining applied to the roof and sidewalls only, the maximum support pressure is at least an order of magnitude lower. [3]6 inch deep I beam weighing 12 lb per foot. [4]8 inch deep I beam weighing 23 lb per foot. [5]12 inch deep wide flange I beam weighing 65 lb per foot. [6]The minimum radius to which I beams can be bent on site is approximately 11 times the section depth. In the case of wide flange beams the minimum radius is approximately 14 times the section depth.

faces. Hence, induced bending will result in support capacities which are lower than those given in Table 9.1. Furthermore, the effect of not closing the support ring, as is normally the case, leads to a drastic reduction in the capacity and stiffness of steel sets and concrete or shotcrete linings. Consequently, the capacities will be lower and the deformations will be larger than those shown in Table 9.1.

9.2.6 *Support interaction example*

In order to illustrate the concepts discussed in the previous sections and to allow the reader to carry out parametric studies of support interaction, a spreadsheet calculation is presented in Figure 9.6. Cell formulae are included in this figure to help the reader to assemble a similar spreadsheet.

Consider the example of a 6 metre diameter shaft ($r_o = 3$ m) excavated in a fair quality, blocky sandstone. The strength characteristics of this rock mass, estimated using the procedures described in Chapter 8, are defined by a cohesion $c = 2.6$ MPa and friction angle $\phi = 30°$. The in situ stress $p_o = 10$ MPa.

As shown in Figure 9.6, failure of the rock mass surrounding the shaft commences when the support pressure p_i is less than the critical pressure $p_{cr} = 2.75$ MPa. The plastic zone radius $r_p = 3.8$ m when the support pressure is zero. The maximum wall displacement without support is $u_i = 47$ mm.

The support selected for this example consists of 34 mm diameter mechanically anchored rockbolts. From Table 9.1, the maximum support pressure $p_{sm} = 0.34$ MPa and the maximum elastic displacement which can be withstood by these bolts is $u_{sm} = 21$ mm.

Note that, in calculating these support characteristics, it has been assumed that the bolt length is equal to 1/3 of the opening diameter and the bolt spacing is assumed to be one half the bolt length. Figure 9.6 shows the load displacement curve for the rockbolt support system, assuming an initial displacement $u_{so} = 25$ mm. This curve intersects the opening displacement curve at a support pressure value of about 0.3 MPa and at a displacement of approximately 43 mm.

It will be evident from this example that even relatively heavy support cannot provide sufficient pressure to prevent the development of a failure zone. In this case it would be necessary to provide a support pressure equal to 2.75 MPa (the value of the critical pressure p_{cr}) in order to prevent this failure and, as can be seen from Table 9.1, this is not available from any support system which can be installed in a reasonable time.

9.3 The PHASES program

The earliest analysis of the elasto-plastic stress distribution around a cylindrical opening was published by Terzaghi (1925) but this solution did not include a consideration of support interaction. Fenner (1938) published the first attempt to determine support pressures for a tunnel in a rock mass in which elasto-plastic failure occurs. Brown

Support interaction analysis for a circular opening in elastic-perfectly plastic rock

Input parameters for rock mass			***Calculated values for rock mass***		
Friction angle	phi=	30 deg	Uniaxial strength	scm=	9.01 MPa
Cohesive strength	coh=	2.6 MPa	Ratio k	k=	3.00
Young's modulus	E=	1000 MPa			
Poisson's ratio	mu=	0.25			

Input parameters for tunnel and in situ stress			***Calculated value for tunnel***		
Radius of opening	ro=	3 m	Critical pressure	pcr=	2.75 MPa
Hydrostatic stress	po=	10 MPa			

Input parameters for support system

Initial deformation before support is installed and effective	uso =	25 mm
Maximum elastic displacement of support (From Table 9.1)	usm =	21 mm
Maximum pressure provided by support (From Table 9.1)	psm =	0.34 MPa

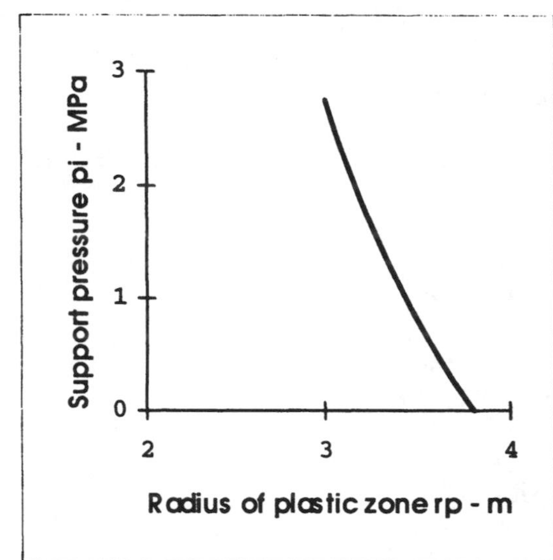

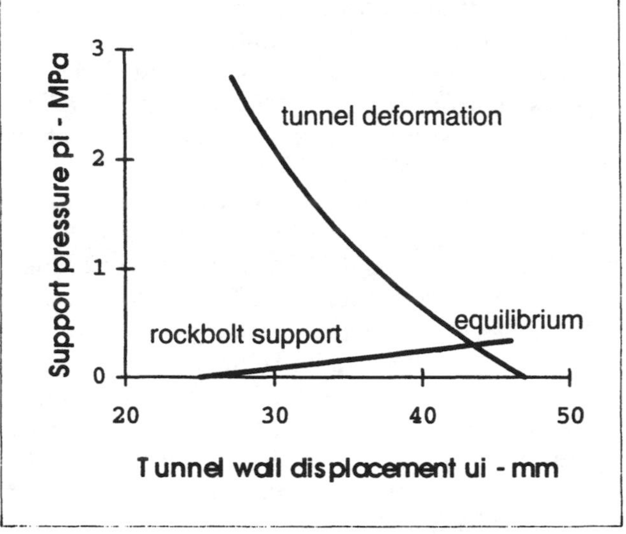

Support pressure pi-MPa	Plastic radius rp- m	Sidewall displace ui - mm
0.000	3.81	47
0.275	3.70	44
0.550	3.59	41
0.825	3.50	38
1.100	3.41	36
1.374	3.33	34
1.649	3.26	32
1.924	3.19	31
2.199	3.12	30
2.474	3.06	28
2.749	3.00	27

Cell formulae

scm=2*coh*cos(phi*pi()/180)/(1 - sin(phi*pi()/180))
k=(1+sin(phi*pi()/180))/(1-sin(phi*pi()/180))
pcr=(2*po-scm)/(k+1)

for pi, starting at zero, add (0.1*pcr) to each subsequent cell up to a maximum of pcr

rp=IF(pi<pcr THEN
ro*(2*(po*(k-1)+scm)/((1+k)*((k-1)*pi+scm)))^(1/(k-1))
ELSE ro)

ui=IF(rp>ro THEN
1000*ro*((1+mu)/E)*(2*(1-mu)*(po-pcr)*((rp/ro)^2)-(1-2*mu)*(po-pi))
ELSE 1000*ro*(1+mu)*(po-pi)/E)

Figure 9.6: Printout of a spreadsheet which can be used for parametric studies of rock support interaction for a circular shaft.

et al. (1983) and Duncan Fama (1993) have reviewed several of the analytical solutions which have been published since 1938. The major difference between these solutions lies in the assumed post-failure characteristics of the rock mass surrounding the tunnel. All the solutions are restricted to the case of a cylindrical opening in a rock mass subjected to a hydrostatic stress field.

The stress field in the rock surrounding most mining excavations is not hydrostatic and very few of these excavations are circular in shape. Consequently, practical applications of the analytical solutions discussed above are severely limited. The main value of these solutions is the understanding of the basic principles of rock support interaction which can be gained from parametric studies involving different material properties, in situ stress levels and support systems.

In order to overcome the limitations of the analytical solution and to provide a tool for practical support design calculations, a program called PHASES, described in Section 7.3.3, on page 81, was developed at the University of Toronto. This program uses a two-dimensional hybrid finite element/boundary element model which is associated with easy to use graphical pre- and post-processors. The graded finite element mesh, which is generated automatically in the pre-processor, surrounds the opening and extends out to the boundary element interface. The use of finite elements in the rock mass immediately surrounding the opening allows for the inclusion of a variety of material types and support systems in the model.

A number of successive excavation stages can be considered and the progressive failure of the rock mass and the reaction of the support system can be tracked for all of these stages. The boundary element model, which surrounds the central finite element model, extends out to infinity. It has the advantages that no additional discretization of this model is required, and that the far-field in situ stresses can be applied without special consideration of the boundary conditions. The Mohr-Coulomb and Hoek-Brown failure criteria can be used to define the strength of the rock mass. The failure of the rock mass is assumed to involve a reduction in strength from one set of strength parameters to a lower set of strength parameters (elastic-brittle-plastic) with provision for dilatancy (volume change) in the failure zone.

9.3.1 *Support interaction analysis using PHASES*

The example of the circular shaft subjected to a hydrostatic stress field, dealt with analytically in the previous section, can be analysed by means of the program PHASES. The results of such an analysis are presented in Figure 9.7.

The automatically generated finite element mesh surrounding the shaft is shown in Figure 9.7a. The program provides a default setting for the number of elements on the opening boundary, but the user can edit this value if necessary. In this case, 75 elements have been specified for the opening boundary to ensure that a fine mesh is created in order to show the details of the failure zone.

The rock mass surrounding the shaft is assumed to fail from the specified strength parameters (c = 2.6 MPa and ϕ = 30°) to the same strength parameters in an elastic-plastic manner. In other words, no

a: Finite element mesh around the opening and extending to the boundary element surface.

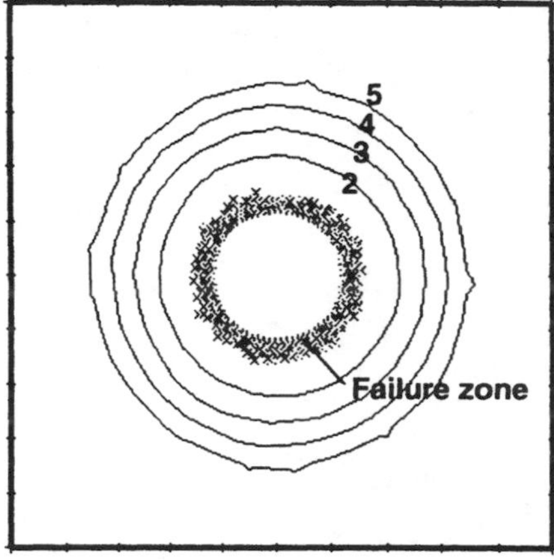

b: Failure zone surrounding the unsupported shaft and contours of the ratio of strength to stress in the elastic rock mass.

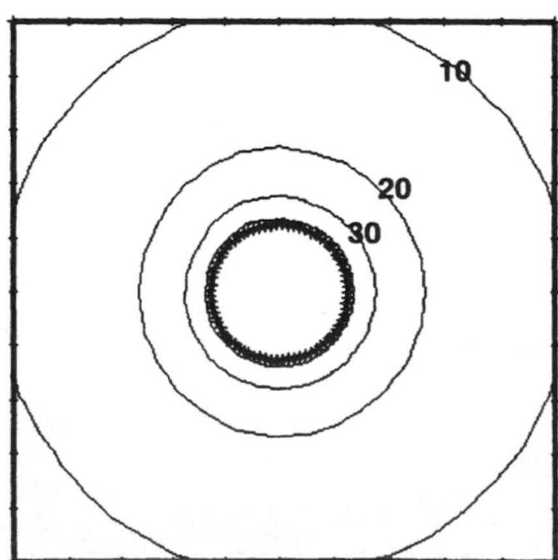

c: Contours of total inward radial displacement in the rock surrounding the unsupported shaft. Maximum displacement of the wall is 46 mm.

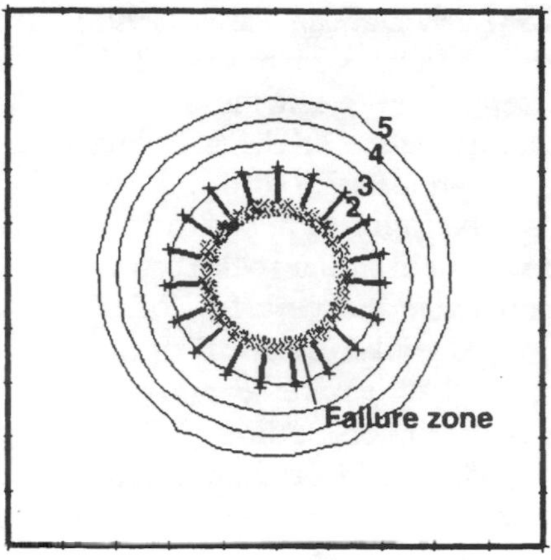

d: Failure zone surrounding the supported shaft and contours of the ratio of strength to stress in the elastic rock mass.

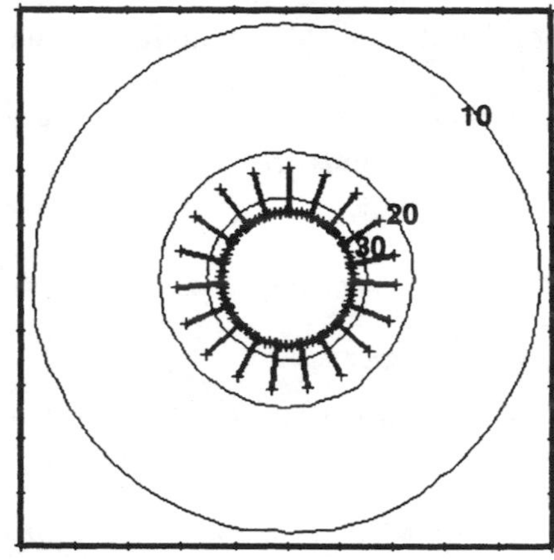

e: Contours of total inward radial displacement in the rock surrounding the supported shaft. Maximum displacement of the wall is 38 mm.

Figure 9.7: Results obtained for an analysis, using the program PHASES, for the circular shaft discussed in Section 9.2.6.

strength drop associated with brittle failure has been allowed in order to ensure that the failure process is the same as that assumed in the analytical model.

The failure zone shown in Figure 9.7b is represented by a series of small crosses, each located at the approximate centre of a triangular element. The failure zone is generated by a succession of calculations in which the excess load, which cannot be carried by a failed element, is transferred onto adjacent elastic elements. If the total load now carried by these elements is too high, they fail, and transfer the excess load onto the next elastic elements. Starting from the excavation boundary where the stresses are highest, failure propagates outwards until the excess load transferred is small enough that it can be carried by the surrounding elastic elements without further failure.

The radius of the failure zone is difficult to measure precisely because the crosses, indicating failure at the boundary of the zone, are located in elements of increasing size. Approximate measurements give a failure zone radius of about 4 m, compared with 3.8 m calculated in the spreadsheet shown in Figure 9.6.

The contours surrounding the failure zone in Figure 9.7b define ratios of available strength to induced stress in the elastic rock mass. The contour, defining the condition where the strength equals the stress, corresponds to the outer boundary of the failure zone.

Figure 9.7c gives contours of displacement in the rock mass surrounding the shaft. These displacements are all radially inward and the maximum boundary displacement is 46 mm, compared with the predicted value of 47 mm in Figure 9.6.

Figure 9.7d shows the failure zone and strength/stress contours for a model in which a radial pattern of 3 m long 34 mm diameter mechanically anchored bolts have been installed. PHASES allows for pre-tensioning of the bolts and also for a specified amount of failure to occur before the bolts become fully effective, corresponding approximately to the delay defined by u_{so} in Figure 9.5. These values have been estimated so that the support pressure is approximately equal to the value of the support pressure of about 0.3 MPa at which equilibrium occurs in Figure 9.6. Note that the radius of the failure zone has been slightly reduced (comparing Figures 9.7d and b) and that the maximum wall displacement has been reduced to 38 mm compared to the 43 mm predicted in Figure 9.6.

The accuracy of these comparisons is of no practical significance since this particular example has been presented to demonstrate some of the capabilities of PHASES and to show its relationship to the analytical model discussed earlier. The examples, included in the program manual and later in this book, demonstrate the use of the powerful capabilities which are included in the program.

10 Progressive spalling in massive brittle rock

10.1 Introduction

One of the problems which is encountered in mining and civil engineering tunnels is slabbing or spalling from the roof and sidewalls. This can take the form of popping, in which dinner plate shaped slabs of rock can detach themselves from the walls with an audible sound, or gradual spalling where the rock slabs progressively, and fall away from the roof and floor. In extreme cases the spalling may be severe enough to be classed as a rockburst.

In all cases the rock surrounding the excavations is brittle and massive. In this context massive means that there are very few discontinuities such as joints or, alternatively, the spacing between the discontinuities is of the same order of magnitude as the dimensions of the opening.

This chapter presents a method for estimating the extent of this slabbing or spalling process in order to provide a basis for the design of rock support. No attempt was made to investigate the detailed rock physics of the process because the aim was to produce a solution, which could be applied by engineers working in the field with minimal data at their disposal. However, a great deal of valuable background information was extracted from the work of a number of authors such as Bieniawski (1967), Cook (1965), Ewy and Cook (1990), Fairhurst and Cook (1966), Hoek (1965), Kemeny and Cook (1987), Lajtai and Lajtai (1975), Martin (1993), Pelli et al. (1991), Zheng et al. (1989), Ortlepp and Gay (1984) and Ortlepp (1992, 1993).

10.2 Examples of spalling in underground excavations

Figure 10.1 shows the typical sidewall spalling which can be observed in boreholes and bored raises in highly stressed rock. The spalling initiates on the hole boundary at points where the tangential compressive stress is highest. These points occur at the intersection of the *minor* principal stress axis and the hole boundary.

Figure 10.2 illustrates spalling in massive quartzite at a depth of about 1,500 m in an underground mine. This spalling occurred over a number of years and did not pose a major threat to the stability of the opening or to the miners.

A more serious situation is illustrated in Figure 10.3 which shows a rockfall caused by spalling in the upper left-hand corner (just above the man's head) of an opening in a large metal mine. Spalling, parallel to the right-hand sidewall of the opening, is also visible in the foreground of the picture. Spalls of this type are relatively uncommon, but they can be very dangerous due to the size of the pieces which can fall from the opening roof.

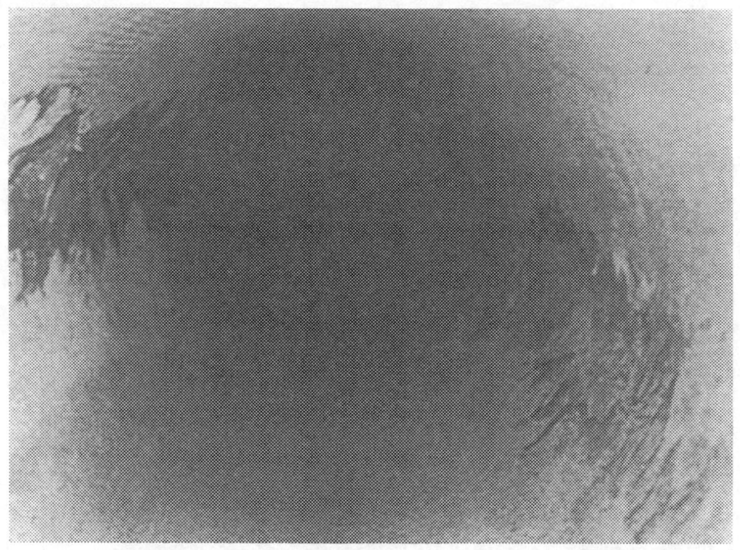

Figure 10.1: Spalling in the sidewall of a bored raise in massive rock. The direction of the major principal stress is shown. Spalling initiates at points of maximum compressive stress concentration which occur at right angles to the major principal stress direction.

Figure 10.2: Spalling of one wall of a drive in quartzite at a depth of about 1,500 m in a uranium mine. In this case, an open stope on the left of the picture caused the stresses in the left-hand wall of the drift to be high enough to initiate spalling. This spalling is relatively minor and is parallel to the drift wall.

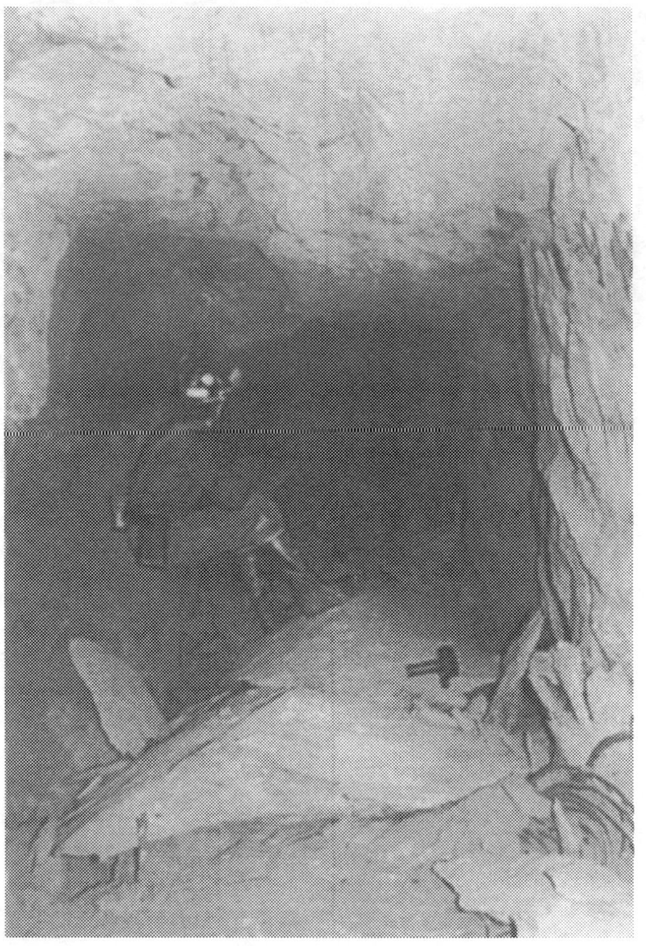

Figure 10.3: Spalling from the roof of an opening in highly stressed rock. The slabs on the floor have come from the left-hand side of the roof, just above the man's head. Spalling parallel to the right-hand sidewall is also visible.

10.3 The AECL Underground Research Laboratory.

Some of the best observations and measurements of spalling and slabbing in massive rock around underground excavations have been carried out in Atomic Energy of Canada's Underground Research Laboratory at Pinawa in Manitoba. Detailed descriptions of the URL and of the observations are contained in publications by Martin and Simmons (1992) and Martin (1990, 1993).

The URL is located within the Lac du Bonnet granite batholith which is considered to be representative of many granite intrusions of the Precambrian Canadian shield. At the 420 level, this granite is massive and almost completely devoid of structural features.

The general layout of excavations at the 420 level in the URL is illustrated in Figure 10.4. The three sites which will be discussed on the following pages are Room 405, Room 413 and the Test Tunnel.

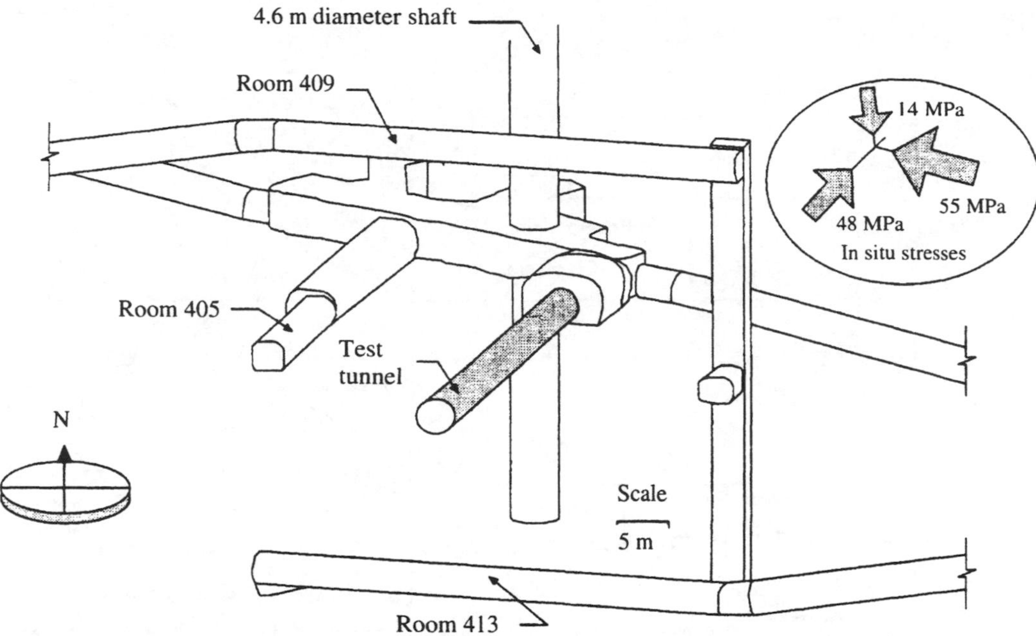

Figure 10.4: Layout of the excavations on the 420 Level of the URL and the locations and profiles used in the analyses presented on the following pages. After Martin (1993).

10.3.1 *In situ stresses at 420 level*

Very extensive studies on the in situ stresses in the rock mass have been carried out at the URL, using a variety of stress measuring techniques (Martin, 1990). These stresses are probably better defined than those at any other site in the world.

The in situ stresses at the 420 level are shown in Figure 10.4. They are:

$\sigma_1 = 55$ Mpa	Parallel to Room 413
$\sigma_2 = 48$ Mpa	Parallel to room 405
$\sigma_3 = 14$ Mpa	Sub-vertical
$\theta = 14°$	Inclination of σ_1 to the horizontal

10.3.2 *Properties of Lac du Bonnet granite*

The properties of the Lac du Bonnet granite have been studied by Lajtai at the University of Manitoba (Lajtai, 1982) and by the CANMET Mining Research Laboratory in Ottawa (Lau and Gorski, 1991).

The triaxial tests on Lac du Bonnet granite from near surface (0-200 m) and from the 420 m level produce very different results. This difference has been attributed to geologically induced damage in the highly stressed rock at the 420 level. In the analyses which follow, it has been assumed that the properties of the intact granite are represented by the results of the tests on the near surface rock ($\sigma_c = 210$ MPa, $m = 28.9$ and $s = 1$) while the m and s values obtained from the tests on samples taken from the 420 level ($\sigma_c = 210$ MPa, $m = 10.84$, $s = 0.296$) are representative of rock of lower quality around the openings.

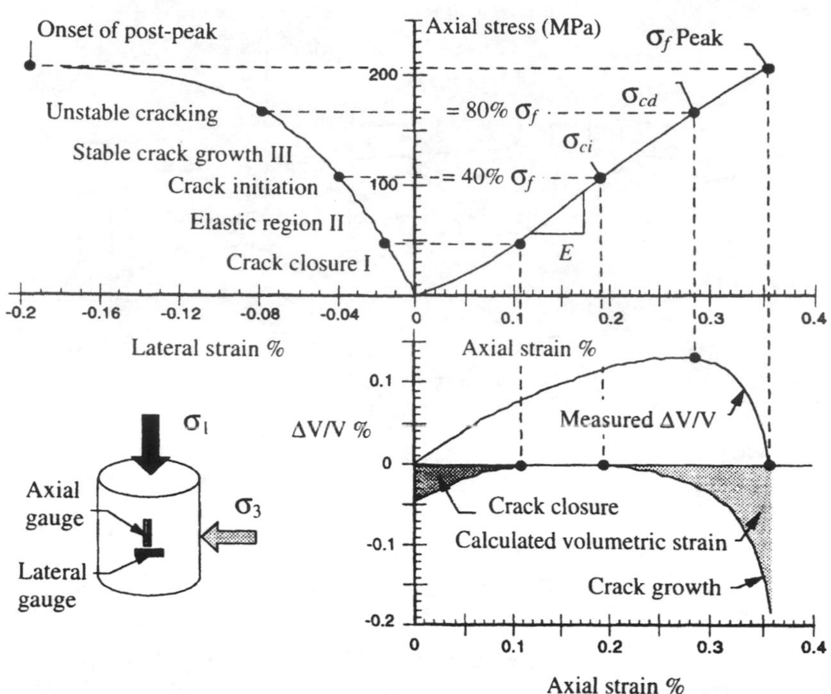

Figure 10.5: Typical stress/strain curves for Lac du Bonnet granite showing points defining peak strength, the onset of unstable crack growth and the onset of stable crack growth. After Martin (1993).

A typical stress/strain curve for Lac du Bonnet granite is reproduced in Figure 10.5 and a full set of triaxial test results for the 420 level granite are given in Figure 10.6.

It has been argued by authors such as Bieniawski (1967) that the long term strength of massive rock is defined by the stress level at which unstable crack growth occurs. As shown in Figure 10.5, this is defined by the peak of the volumetric strain curve which occurs at between 70 and 80% of the peak strength of the rock. The values defining the onset of unstable crack growth will be used in the analysis of progressive failure around the various openings considered.

10.3.3 *URL Rooms 413 and 405*

The main difference between these two excavations, as shown in Figure 10.4, is that room 413 is aligned parallel to the maximum principal in situ stress ($\sigma_1 = 55$ MPa) while room 405 is aligned parallel to the intermediate principal stress ($\sigma_2 = 48$ MPa). As shown in Figure 10.7, very little spalling was observed on the boundary of room 413 (left-hand figure) while quite severe spalling occurred in the upper left-hand part of the roof and in the floor of room 403 (right-hand figure).

Failure zones, predicted by the program PHASES, for rooms 413 and 405 are illustrated in the lower drawings in Figure 10.7. The PHASES models for these excavations were identical except for the applied stress conditions. The excavation shapes were constructed by tracing drawings presented by Martin (1993). The properties assigned to the granite were as follows:

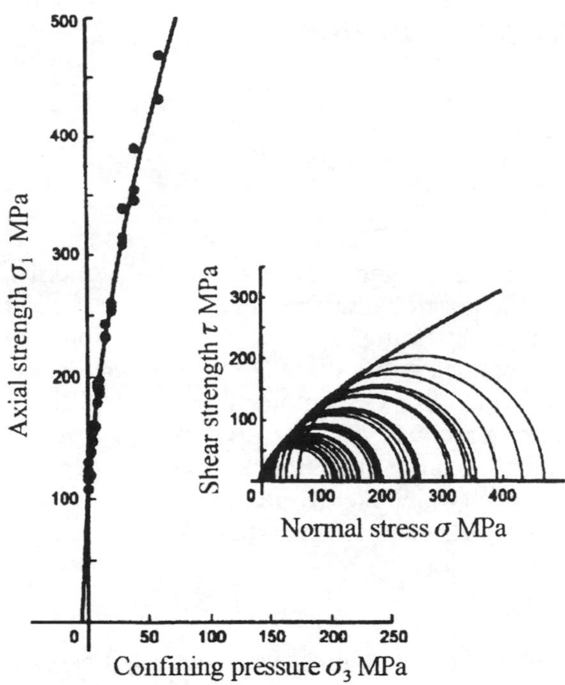

Figure 10.6: Results of triaxial tests on granite samples from the URL 420 level. Stresses are for the onset of unstable crack growth as defined in Figure 10.5. The fitted curved are defined by the following values: σ_c = 210 MPa, m = 10.84 and s = 0.296.

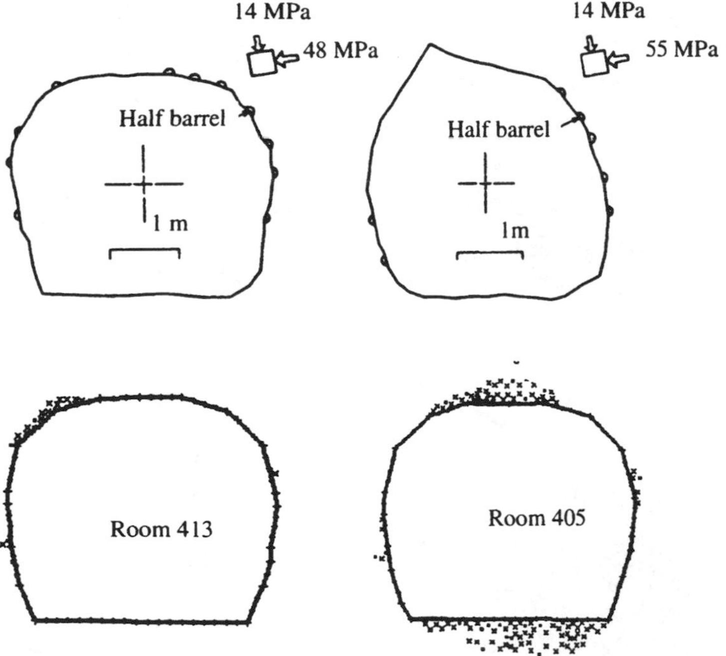

Figure 10.7: Comparison between observed and predicted failure in rooms 413 and 405 on the 410 level in the URL. The upper illustrations are observed conditions in room 413 and 405 respectively. The lower drawings show failure zones predicted by the program PHASES, applying the in situ stress conditions shown.

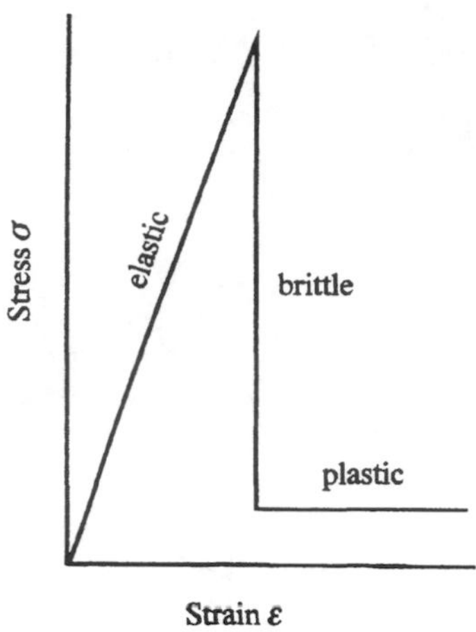

Elastic-brittle-plastic behaviour assumed for massive brittle rock

Young's modulus	$E = 60{,}000$ MPa
Poisson's ratio	$\nu = 0.2$
Intact strength	$\sigma_c = 210$ MPa
undisturbed granite	$m = 10.84$
undisturbed granite	$s = 0.296$
failed granite	$m_r = 1.0$
failed granite	$s_r = 0.01$

Note that the failed granite is assigned very low strengths in order to simulate the elastic-brittle-plastic failure process which results in the rock spalling and falling away from the roof of the excavation.

The failure zones predicted by the PHASES model show a reasonable similarity to the observed failure. It was found that the shape of the final failure zone was very sensitive to the shape of the excavation from which failure initiated, but that the depth and volume of the failure zone (which are of greater interest for support design) were controlled by the material strength and in situ stresses.

10.3.4 *URL Test Tunnel*

As shown in Figure 10.4, the Test Tunnel was excavated parallel to Room 405. The 3.5 m diameter test tunnel had a circular profile and was excavated in 1 m and 0.5 m increments using perimeter line drilling and mechanical breaking of the rock stub. Excavation of each increment could be completed in two 8 hour shifts, but experimental activities constrained progress to one round about every three days.

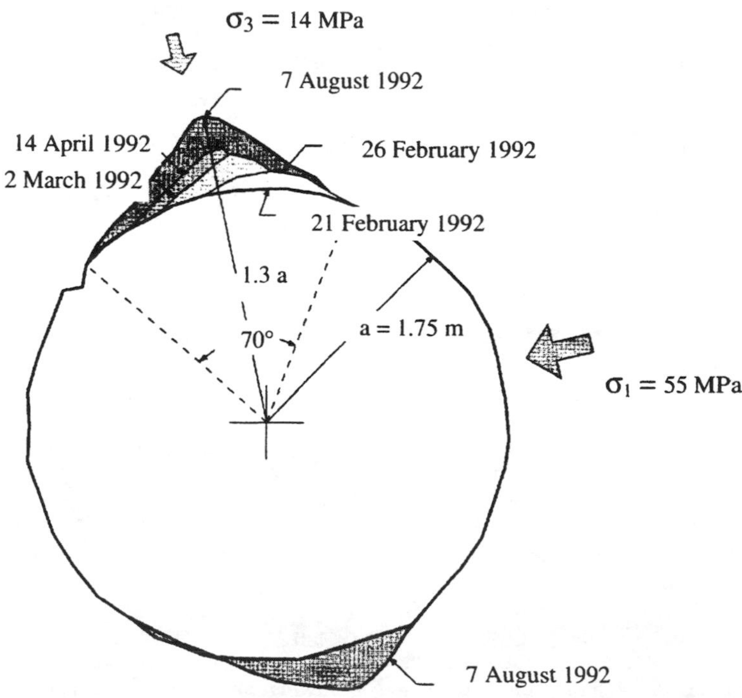

Figure 10.8: Progressive development of the notch geometry in the roof and floor of the URL Test Tunnel over a five month period.

Figure 10.9: PHASES model showing the predicted shape of the failure zone for the Test Tunnel.

The tunnel was excavated over a six month period. Failure in the roof and floor was observed immediately as each excavation round was taken, and progressed as the test tunnel was advanced. Figure 10.8 illustrates the development of the notch in the roof over about a five month period. The progressive development of the notch in the floor is not available because the floor always contained 'tunnel muck' until the tunnel advance was completed. However, the final shape of the notch in the floor is very similar to the notch in the roof. The dates given in Figure 10.8 do not reflect the actual times required for the notch to develop, but the dates of the actual notch survey. The thickness of the spalling slabs, which created the notch, varied from a few millimetres to tens of millimetres and there did not appear to be any preferred direction of slabbing, i.e., the slabs formed on both sides of the notch. Regardless of the process causing the notch development, the orientation and geometry of the notch was consistent from the start of the test tunnel to the end of the test tunnel, and this orientation is consistent with the 14 degree plunge of the major principal stress.

The failure zones predicted by the model PHASES, using identical input to that used for the analysis of Room 405, are illustrated in Figure 10.9. The correspondence between the observed and predicted failure is considered acceptable for most practical support design purposes.

10.4 Example from El Teniente Mine, Chile

The following example is taken from an unpublished research report by P.K. Kaiser and it is included in this chapter with permission from the El Teniente Mine in Chile. The tunnel was excavated in Andesite for which a Rock Mass Rating of *RMR* = 62 to 69 was estimated. Using the procedures described in Chapter 8, the rock mass strength is estimated as:

Uniaxial compressive strength of intact rock	$\sigma_c = 150$ MPa
Hoek Brown constant for undisturbed rock	$m = 8.35$
Hoek Brown constant for undisturbed rock	$s = 0.032$
Deformation modulus	$E = 25{,}000$ MPa
Poisson's ratio	$\nu = 0.3$
Failed Andesite	$m_r = 1.0$
Failed Andesite	$s_r = 0.01$

The in situ stresses for this example were assumed, from field measurements, to be $\sigma_1 = 38$ MPa, $\sigma_2 = 31$ MPa, $\sigma_3 = 24$ MPa and the inclination to the horizontal $\theta = 28°$.

As for the other examples discussed in this chapter, the coincidence between the observed and predicted failure zones shown in Figure 10.10 is considered adequate for most practical applications. The predicted results were obtained by using all of the default settings of the program PHASES and the material strengths and in situ stresses defined above. Again, a massive brittle stress drop has been used (defined by the failed Andesite properties of $m = 1$ and $s = 0.01$) to simulate spalling of the roof and sidewalls. Some failure of the floor of the actual tunnel has occurred as suggested by the predicted failure zone in Figure 10.10.

10.5 South African experience

Ortlepp and Gay (1984) published details of an experimental tunnel at a depth of 3,250 m below surface in massive quartzite in the East Rand Proprietary Mine in South Africa. The tunnel was subjected to significant stress changes from 1975 to 1980 as a result of mining of adjacent stopes. Severe spalling occurred during this time, resulting in the final shape illustrated in Figures 10.11 and 10.12. The excavated tunnel width was 1.5 m and the final 'overbreak' extended to a span of about 4 m.

From the details on intact rock contained in the paper, the rock mass strength was estimated for an *RMR* value of 75, allowing for blast damage and a few structural features. The rock mass strength used in the PHASES analysis was as follows:

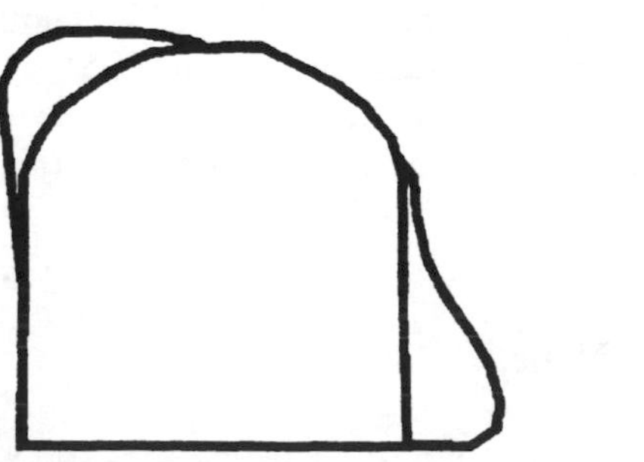

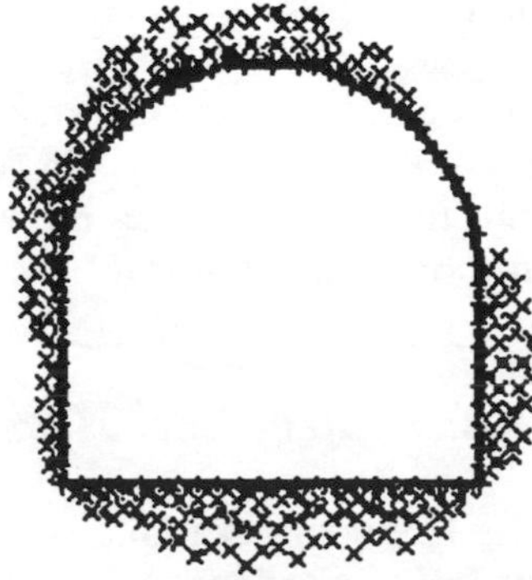

Figure 10.10: Comparison between observed and predicted failure around a tunnel in the El Teniente mine in Chile.

Uniaxial compressive strength of intact rock	$\sigma_c = 350$ MPa
Hoek Brown constant for undisturbed rock	$m = 9.42$
Hoek Brown constant for undisturbed rock	$s = 0.062$
Deformation modulus	$E = 40{,}000$ MPa
Poisson's ratio	$\nu = 0.2$
Failed quartzite	$m_r = 1.0$
Failed quartzite	$s_r = 0.01$

The in situ stresses, from the figures published in the paper, were estimated to be σ_1 = 225 MPa, σ_2 = 85 MPa, σ_3 = 220 MPa. The major principal stress is 10° off vertical.

The results of a PHASES analysis of this problem are reproduced in Figure 10.13 and the failed material is shown by the × marks. Compared with the surveyed 'overbreak' profile reproduced in Figure 10.12, it is evident that the analysis has over-estimated the extent of the failure. The location of the failure zones in the sidewalls coincides with the descriptions in Ortlepp and Gay's paper. The predicted 'wing' cracks, propagating in the direction of the major principal stress, suggests that the estimated strength is too low (Hoek, 1965). However, these cracks would be difficult to detect underground. Therefore, it is not known whether or not they existed. It is also not known whether failure extended into the floor as suggested by the analysis.

In setting up this analysis, an arbitrary decision was made to reduce the laboratory strength values of the quartzite (σ_c = 350 MPa, m = 23 and s = 1) to correspond to the properties of a 'blocky/good'

Figure 10.11: Overbreak resulting from spalling in a tunnel in massive quartzite at a depth of 3250 m below surface in the East Rand Proprietary Mine (ERPM) in South Africa. Photograph provided by Mr David Ortlepp.

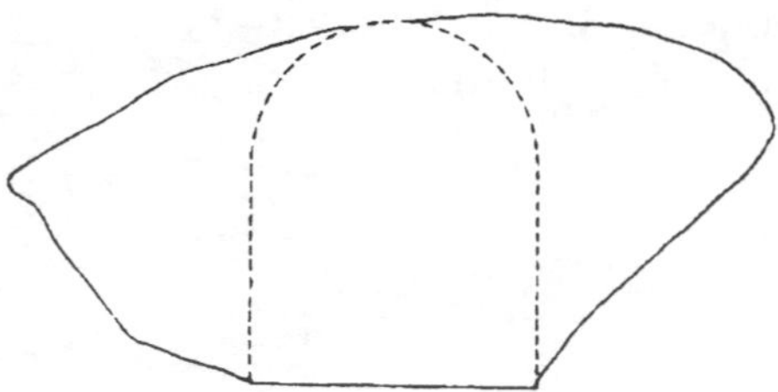

Figure 10.12: Surveyed overbreak in the rock surrounding the tunnel at benchmark 7/8. After Ortlepp and Gay (1984).

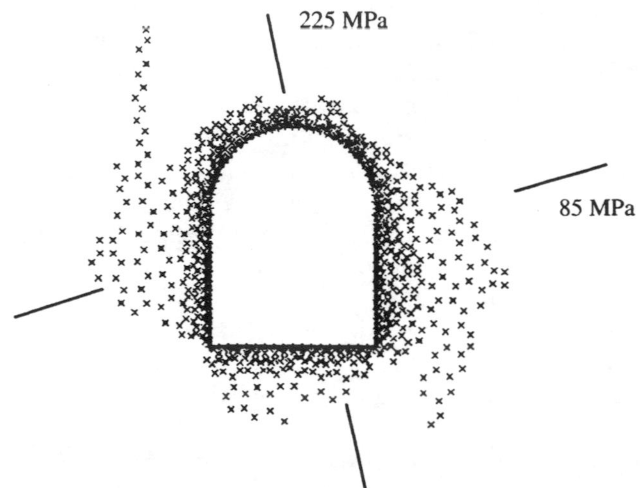

Figure 10.13: Zone of failure predicted by a PHASES analysis.

rock mass (see Table 8.4) with an *RMR* value of 75, giving $\sigma_c = 350$ MPa, $m = 9.42$ and $s = 0.062$. A check, using properties estimated for an *RMR* value of 80, resulted in an under-estimation of the extent of the failure zone and an elimination of the 'wing' cracks. The surveyed profile shown in Figure 10.12 appears to lie between these two predictions.

This analysis suggests that the extent of the failure zone is very sensitive to the assumed rock mass properties. It is very unlikely that better estimates of rock mass strength, than those used in this analysis, are likely to be available in the near future. Consequently, some inaccuracy in the size and shape of the predicted failure zone must be anticipated. This difference is of academic rather than practical significance since available support design techniques are not sufficiently refined to take these differences into account.

A second example of a highly stressed tunnel in a South African gold mine has been described by Ortlepp (1993). This tunnel was mined in massive quartzite at a depth of 2,700 m below surface and severe spalling occurred in the upper right-hand corner of the tunnel as illustrated in Figure 10.14. Slabbing of the left-hand sidewall is also known to have occurred but the extent of this slabbing is not clear.

Figure 10.14: Spalling around a tunnel in massive quartzite at a depth of 2,700 m below surface in a South African gold mine. Photograph provided by Mr David Ortlepp.

No information on the material properties is available. Hence, the properties have been estimated from published information on Witwatersrand quartzite (Hoek, 1965). The in situ properties are based upon the assumption that the quartzite is 'blocky' and 'good' to 'very good' with an *RMR* value of 80. The rock mass properties used in the PHASES model are defined by:

Uniaxial compressive strength of intact rock	$\sigma_c = 200$ MPa
Hoek Brown constant for undisturbed rock	$m = 16$
Hoek Brown constant for undisturbed rock	$s = 0.33$
Deformation modulus	$E = 90{,}000$ MPa
Poisson's ratio	$\nu = 0.2$
Failed quartzite	$m_r = 1.0$
Failed quartzite	$s_r = 0.01$

In situ stress values have been derived from a number of stress measurements in the general area of this tunnel and are as follows:

$\sigma_1 = 90$ MPa	Sub-vertical
$\sigma_2 = 89$ Mpa	Parallel to tunnel
$\sigma_3 = 70$ Mpa	Sub-horizontal
$\theta = 50°$	Inclination of σ_1 to the horizontal.

The failure zones predicted by the PHASES model study are shown in Figure 10.15 and appear to agree well with the type of damage visible in the photograph reproduced in Figure 10.14.

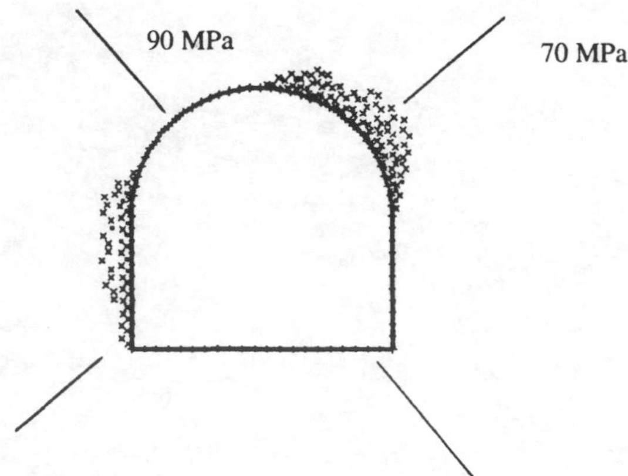

Figure 10.15: Failure zones predicted by PHASES for the second tunnel example.

Ortlepp, in a personal communication, has commented that the PHASES model does not reproduce the tensile slabbing process which is apparent in the upper right-hand corner of the tunnel illustrated in Figure 10.14. Figure 10.15 shows that shear failure (denoted by the × symbols) is the dominant mode of failure predicted for this region. While this difference is not disputed, the reduction of confinement as a result of the progressive failure process (controlled by the large brittle stress drop) is believed to produce a similar end result. Hence, until a progressive tensile slabbing model is available, the use of the PHASES model appears to provide a reasonable tool for the prediction of progressive spalling in massive brittle rock.

10.6 Implications for support design

The spalling process discussed in this chapter tends to start very close to the face of the tunnel and, while the full extent of the failure zone may take time to develop, small rockfalls can occur close to the face and can pose a threat to work crews. Traditional methods of supporting the rock in such cases involve the installation of relatively short mechanically anchored rockbolts with wire mesh fixed under the faceplates. The purpose of this support is to carry the dead weight of the broken rock and to prevent rockfalls close to the working face. It is interesting to consider whether alternative methods of support would provide greater control of the spalling process.

The program PHASES includes many options for support installation and two of these options are investigated for the case of the mine tunnel in Chile, described on page 119. These options are the installation of rockbolts and the application of shotcrete.

10.6.1 *Rockbolting*

Using the PHASES model of the 4 m span tunnel created for the El Teniente example, a pattern of 2 m long 25 mm diameter mechanically anchored rockbolts was installed on a grid of 1 m × 1 m in the arch of the roof and upper sidewalls. The bolts were assigned a capacity of 20 tonnes and tensioned to 10 tonnes after installation. A delay in the activation of the bolts was specified to simulate the fact that they could only be installed about 2 m behind the face. This delay is specified in the program as a percentage of the load re-distribution which is allowed to occur before the bolts are activated and, in this case, a 30% factor was used.

The results of this analysis are presented in Figure 10.16 which, compared with Figure 10.10, shows that the bolts have very little influence upon the failure zone. This finding is not too surprising in view of the high stress levels (>50 MPa) in the rock surrounding the tunnel and the fact that the support pressure generated by the rockbolt pattern is less than 0.3 MPa.

10.6.2 *Shotcrete*

The addition of a layer of 100 mm thick shotcrete was simulated in the PHASES model by placing this against the excavation boundary as a second material. This process implies that the support provided by the shotcrete is activated as soon as the tunnel is excavated. It was also assumed that the strength of the shotcrete would instantly achieve the 7 day strength defined by $\sigma_c = 35$ MPa, $m = 8$ and $s = 1$. A modulus of $E = 20{,}000$ MPa and a Poisson's ratio of $\nu = 0.2$ were assumed for the shotcrete. The failure was assumed to be elastic-perfectly plastic, simulating the post-failure behaviour of steel fibre reinforced shotcrete.

In spite of these very optimistic assumptions on the properties and the action of the shotcrete layer, Figure 10.17 shows that the addition of this support has a minimal influence upon the extent of the failure zone in the rock surrounding the tunnel.

10.6.3 *Discussion*

The results presented in Figures 10.16 and 10.17 suggest that the installation of support will not prevent the onset and propagation of spalling in massive rock surrounding a highly stressed tunnel. This confirms practical experience which suggests that support systems,

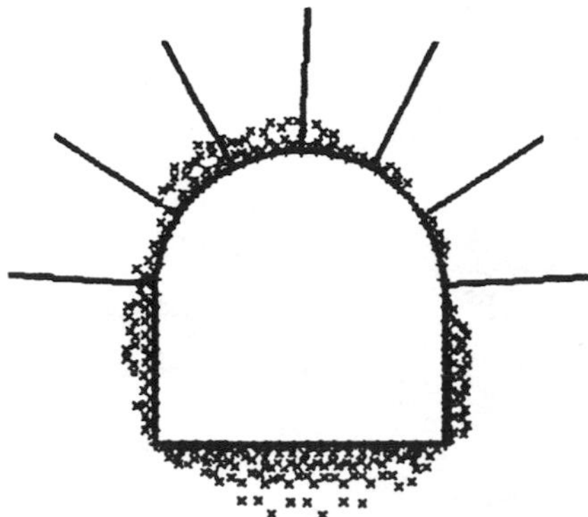

Figure 10.16: Zone of failure in the Andesite surrounding a 4 m span tunnel which has been rockbolted with 2 m long 25 mm diameter rockbolts.

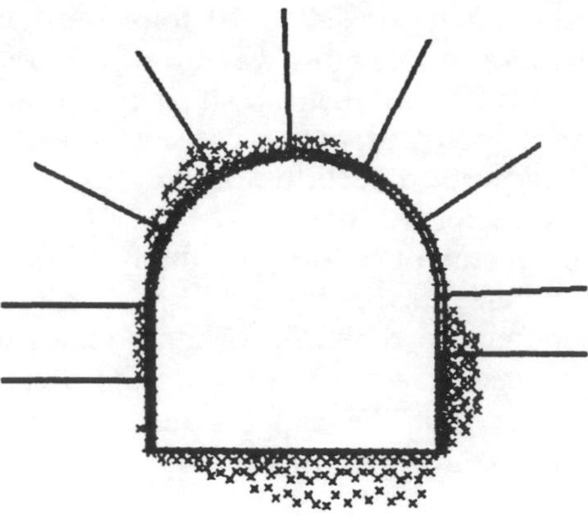

Figure 10.17: Zone of failure in the Andesite surrounding a 4 m span tunnel supported by means of a 100 mm thick layer of steel fibre reinforced shotcrete and a pattern of 2 m long, 25 mm diameter fully grouted and tensioned dowels.

such as rockbolts and shotcrete, do not prevent rock failure from initiating and that their purpose is to control this failure once it has started.

The early installation of support systems which are too stiff will result in overstressing and failure of the support. Consequently, the support should be compliant enough to accommodate the dilation generated by the failure process, but strong enough to support the dead weight of the broken rock. Ungrouted, mechanically anchored rockbolts with wire mesh held under the faceplates are appropriate for small tunnels in which the amount of spalling is limited. Higher capacity multi-strand cables with ungrouted 'stretch' sections (created by sheathing the cable in a plastic sleeve before grouting) should be used for larger excavations or to contain severe spalling.

The application of the PHASES model, in the manner described in this chapter, will give a reasonable assessment of the location and the extent of the zone of potential spalling. The extent of this failure zone can be used to estimate the capacity and the length of support elements such as rockbolts or cables.

11 Typical support applications

11.1 Introduction

The wide variety of orebody shapes and rock mass characteristics which are encountered in underground mining mean that each mine presents a unique design challenge. 'Typical' mining methods have to be modified to fit the peculiarities of each orebody. Similarly, service excavations such as shafts, ramps, haulages and drawpoints have to be engineered to fit in with the geometry of the mine, the sizes of the equipment to be used and the characteristics of the rock mass.

In attempting to present 'typical'support applications, the authors recognise that this is an almost impossible task. Almost every experienced mining engineer who reads this chapter will find that these examples do not fit their own mining conditions very well. Nevertheless, there are a number of fundamental concepts which do apply to support design and these concepts will remain valid, even if the details of the support systems are changed to suit local conditions.

An attempt has been made to capture and describe these fundamental concepts in the following examples.

Each example has been chosen to illustrate the fundamental principles which can be used as a starting point for a support design. In describing the design of typical drawpoint support systems, the factors which control the performance of the support system (abrasion, vibration, secondary blasting damage, stress changes due to stoping) are described and illustrated by means of photographs and sketches. Typical support systems which perform well in drawpoints are considered; reasons for poor performance of other support systems are discussed. The extent to which the reader wishes to use these 'typical' designs or to modify them depends upon the particular circumstances under consideration at the time. In many cases, simple 'rule of thumb' designs are adequate while, in other cases, extensive analysis and redesign may be necessary in order to arrive at an acceptable practical solution. Hopefully, the information contained in the other chapters of this volume will be of assistance in these detailed designs.

11.2 'Safety' support systems

The simplest form of underground excavation support is that which is installed solely for 'safety' reasons. This support is not called upon to carry very heavy loads due to large wedge failures or to massive stress induced instability, but its function is to provide an acceptable level of safety for personnel and equipment in the mine.

Note that there are hundreds of kilometres of mining and civil engineering tunnels around the world which have been successfully mined and operated without support. These tunnels are either in very good quality rock or they are used infrequently enough that safety is not a major issue. The decision on when support is required in such tunnels is a very subjective one, since there are very few guidelines and those which do exist vary widely from country to country. Possibly the only consistent guideline is that heavily trafficked openings, such as shafts, ramps and haulages, should have rockbolts and mesh installed to protect personnel and equipment from rockfalls.

Figure 11.1 illustrates a ramp excavation in an underground mine. The ramp is located in the footwall, some distance from the orebody, and no significant stress induced instability problems were anticipated in this excavation. The rock mass is of relatively good quality with a few joints and blast induced fractures. Most of the loose material was removed by scaling before installation of the support. Under these circumstances, there was clearly no need to design a support system to control slabbing, spalling, stress induced displacements or large wedge failures. The sole purpose of the support was to prevent small rockfalls from injuring personnel and damaging equipment.

The support illustrated in Figure 11.1 consists of a pattern of rockbolts and welded wire mesh extending over the roof and upper sidewalls of the ramp. Since the rockbolts are not required to carry significant loads, 2 m long mechanically anchored rockbolts were installed on a 2 m × 2 m grid with the weld mesh secured under the face plates. The working life of this ramp was only a few years and so corrosion problems were not considered to be a major factor. For excavations requiring more 'permanent' support, the ungrouted rockbolts and welded wire mesh shown in Figure 11.1 would not be an appropriate choice because of the risk of corrosion.

Figure 11.2 shows a conveyor tunnel in which more ‘permanent’ support has been installed. Here, grouted rebar was placed in a pattern in the roof and upper sidewalls of the tunnel and then the entire tunnel surface was covered by a layer of about 50 mm thick shotcrete. This support system is obviously more substantial than that illustrated in Figure 11.1 and it has been designed for a life of about ten years. The expense of this support is justified because very little maintenance or rehabilitation would be required for the life of the tunnel. Such rehabilitation can be very expensive and, in the case of a conveyor tunnel or a similar critical route in the mine, the suspension of operations due to rockfalls would be a serious problem.

The type of instability problems which can occur in an unsupported excavation are illustrated in Figure 11.3. Here the rock mass is relatively closely jointed and, as a result of blasting in adjacent stopes, small wedges and blocks have fallen from the tunnel roof. Figure 11.4 shows a number of ‘typical’ support installations. These can be considered for situations where no significant instability is anticipated but where there is a need to ensure that the opening remains safe for personnel and equipment.

‘Safety’ bolts or dowels would generally not be required to carry a load in excess of about one ton and so very light bolts can be used. Mechanically anchored rockbolts or friction anchored dowels, such as ‘Swellex’ or ‘Split Sets’ are adequate for these installations. The choice of which system to use depends upon cost and availability and upon the ease and speed of installation.

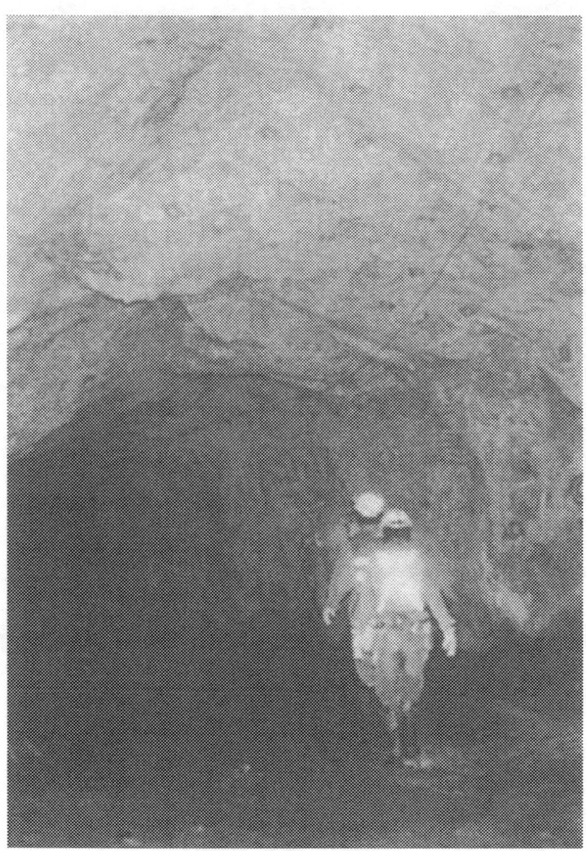

Figure 11.1: Rockbolts and welded wire mesh installed in the roof and upper sidewalls of a ramp excavation in an underground mine. This support has been installed to prevent injury to personnel and damage to equipment from small rockfalls.

Figure 11.2: A conveyor tunnel with grouted rockbolts and shotcrete support. Here the tunnel has been designed for a life of about ten years and possible corrosion of the support system was a major factor in the choice of the support.

Figure 11.3: An accumulation of small blocks and wedges which have fallen from the roof and walls of an unsupported tunnel in a closely jointed rock mass. These small failures, induced by mining activities in adjacent stopes, could be controlled by the installation of light support systems such as those illustrated in Figures 11.1 and 11.2. A layer of shotcrete can provide very effective support for this type of rock mass.

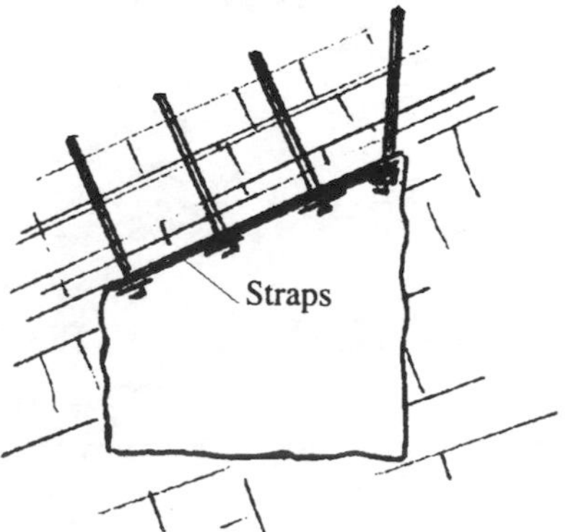

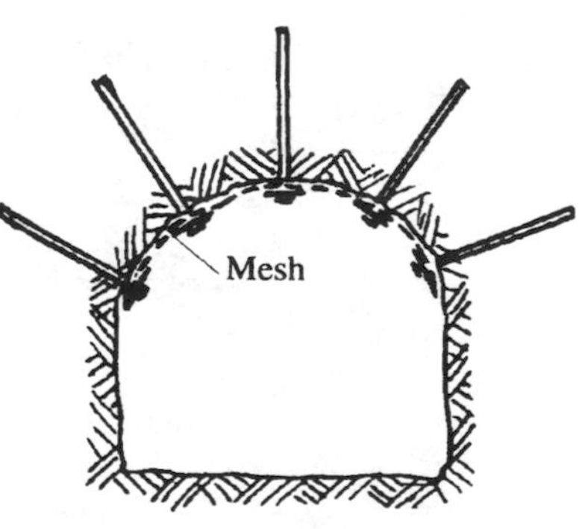

a) Conventional rectangular excavation

b) Arched roof excavation

c) Shanty back excavation in bedded rock

'Typical' rockbolt pattern is 2 m long on a 2 m × 2 m grid

Figure 11.4: Typical support systems which can be installed to improve the safety of service excavations such as haulages, ramps and conveyor tunnels.

Where it is anticipated that corrosion is likely to be a problem, the rockbolts or dowels will either have to have a protective coating applied (usually by the manufacturer) or they would have to be grouted in place. This question is discussed in greater detail in Chapter 12 which deals with rockbolts.

When the rock mass is closely jointed and there is a danger of small blocks and wedges falling out between the rockbolts, wire mesh can be installed behind the rockbolt washers or face plates. Where space is limited or when the rock surface is very rough, chain link mesh is probably the best choice. Where there is sufficient room to work and where the rock surfaces are reasonably smooth, welded wire mesh is a better choice. When there is a possibility that shotcrete will be applied over the mesh, welded wire mesh will result in a better final product than chain link mesh. This is because the chain link mesh obstructs the proper placement of shotcrete and voids are formed where the shotcrete has not been able to penetrate the mesh. This is less of a problem with welded wire mesh, because of the smaller obstruction created by the crossing wires.

Wire mesh should be firmly attached to the rock by washers or face plates on the rockbolts or dowels.

Mesh is not easy to protect against corrosion and, where this is likely to be a serious problem, the replacement of the mesh with steel fibre reinforced shotcrete should be considered. This question is discussed in more detail in Chapter 15 which deals with shotcrete.

Straps can be useful for providing support between rockbolts installed in bedded rock masses. In such cases, the straps should be installed across the 'grain' of the rock as illustrated in the margin photograph. Straps, installed parallel to the strike of significant discontinuities in the rock, will serve little purpose. Where the rock is 'blocky' and where there is no obvious preferred direction for placing the straps, wire mesh should be used instead of straps.

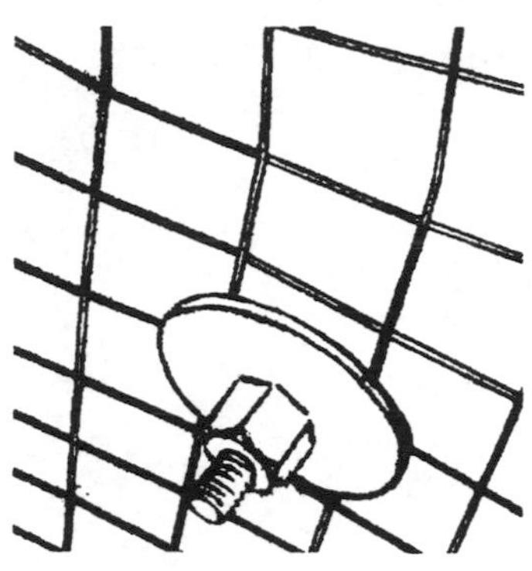

Welded wire mesh is a better choice than chain link mesh where the excavation surfaces are reasonably smooth and where there is enough room to work.

11.3 Permanent mining excavations

Shafts, shaft stations, underground crusher chambers, underground garages and lunch rooms are examples of 'permanent' mining excavations. Because of the frequent use of such excavations by mine personnel and because of the high capital cost of the equipment housed in these excavations, a significantly higher degree of security is required than for other mine openings.

As for the case of 'safety' support systems, discussed in the previous section, security rather than stability is generally the main factor which has to be taken into account in the design of the support systems. These excavations are usually designed for an operational life of tens of years. Consequently, corrosion is a problem which cannot be ignored. In some cases, galvanised or stainless steel rockbolts have been used in an attempt to control corrosion problems. However, fully grouted dowels, rockbolts or cables are usually more effective and economical. Fibre or mesh-reinforced shotcrete, rather than mesh or straps, is used on exposed surfaces and, in many cases, the thickness of the shotcrete may be of the order of 100 to 150 mm.

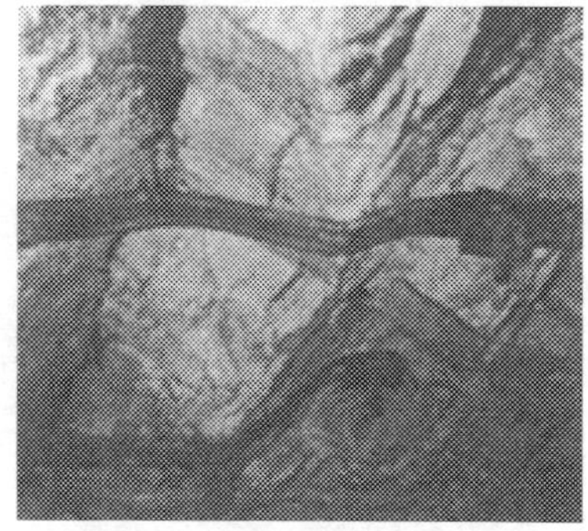

Straps, when used, should be placed across the 'grain' of the rock as shown

Figure 11.5: Partially completed excavation for an underground machine shop. Exceptional care has been taken with the blasting to ensure that there is no excessive overbreak and that the rock surfaces are as smooth as possible. A pattern of grouted rockbolts has been installed in the roof and upper sidewalls. A layer of shotcrete will be applied to the exposed surfaces of the excavation to secure small pieces of rock which could fall from between the rockbolts. This shotcrete will also improve the appearance of the walls and roof and provide a better background for lighting.

11.4 Drawpoints and orepasses

Drawpoints and orepasses require special consideration in terms of support design. These openings are generally excavated in undisturbed rock. Consequently mining is relatively easy and little support is required to stabilise the openings themselves. Once mining starts and the drawpoints and orepasses are brought into operation, the conditions are changed dramatically and serious instability can occur if support has not been installed in anticipation of these changes.

Abrasion, due to the passage of hundreds of tonnes of broken ore, can pluck at loose rock on the opening surfaces and can cause progressive ravelling and eventual collapse. Stress changes, due to the mining of adjacent or overlying stopes, can result in failure of support. Secondary blasting of hang-ups in the drawpoints or orepasses can cause serious damage to the surrounding rock. In other words, the rock surrounding these openings requires considerable assistance if it is to remain in place for the working life of the opening.

A typical drawpoint in a large metal mine

Failure of the brow of a drawpoint can cause loss of control of the broken rock in the stope resulting in serious dilution problems. Figure 11.6 shows a drawpoint which collapsed and where most of the ore in the stope had to be abandoned. In such cases, there is considerable economic incentive to install the correct reinforcement during development of the openings in order to avoid costly remedial work later.

Figure 11.7 shows a drawpoint, which was successfully reinforced by means of untensioned cement grouted reinforcing bars that were installed during development of the drawpoint. As shown in Figure 11.8, the 3 m long rebars were grouted into the rock above the brow of the drawpoint, from the drawpoint and from the trough drive, before the final blast of the brow area was carried out. This means that the rock mass was pre-reinforced and that the individual pieces in the rock mass were kept tightly interlocked throughout the operating life of the drawpoint. Plain rebars, with no face plates or end fixings, were used so that movement of the ore through the drawpoint would not be obstructed and so that the faceplates would not be ripped off, as would happen if mechanically anchored bolts were used.

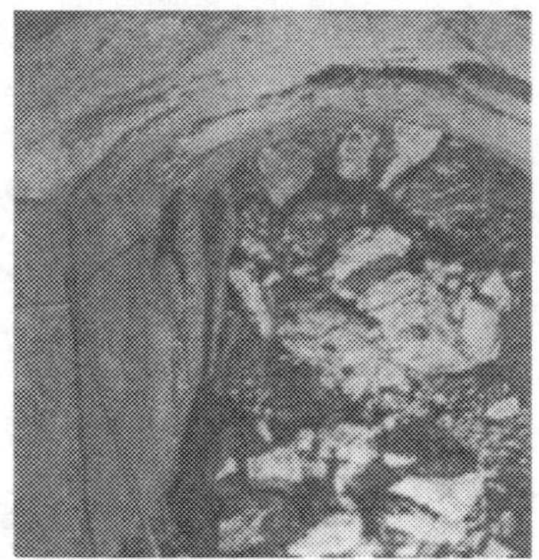

Attempting to reinforce a drawpoint with concrete or steel sets after failure has started in unlikely to be successful

In general, attachments should not be used on the ends of the reinforcement exposed in the drawpoint brow area. Faceplates, straps or mesh will tend to be ripped off and may pull the reinforcement with them. Similarly, surface coatings such as shotcrete should only be used where the surrounding rock is clean and of high quality and where the drawpoint is only expected to perform light duty.

Grouted rebar is a good choice for drawpoint reinforcement in cases where the rock is hard, strong and massive. When the rock is closely jointed and there is the possibility of a considerable amount of inter-block movement during operation of the drawpoint, rebar may be too stiff and the rock will break away around the rebar. In such cases, the use of grouted birdcage or nutcage cables (described in Chapter 13) should be considered. These cables are flexible and have a high load carrying capacity as a result of the penetration of the grout into each of the 'cages' in the cable.

Shotcreting the exposed rock surrounding a drawpoint may be successful provided that the rock mass is sound and that only small tonnages are to be drawn

The design of support for orepasses is similar to that for drawpoints, except that access to install the support is generally not as simple as for drawpoints. In addition, an orepass is required to handle much larger tonnages of ore and may be required to remain in operation for many years.

Figure 11.6: Failure of the brow of a drawpoint resulting in the loss of the ore remaining in the stope.

Figure 11.7: Drawpoint reinforced with cement grouted untensioned rebar, installed during development of the drawpoint.

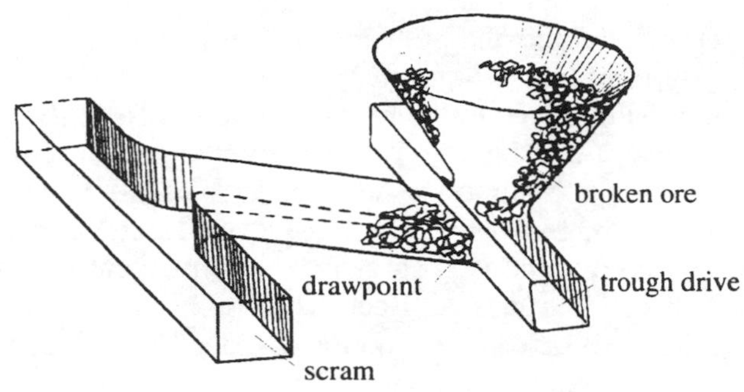

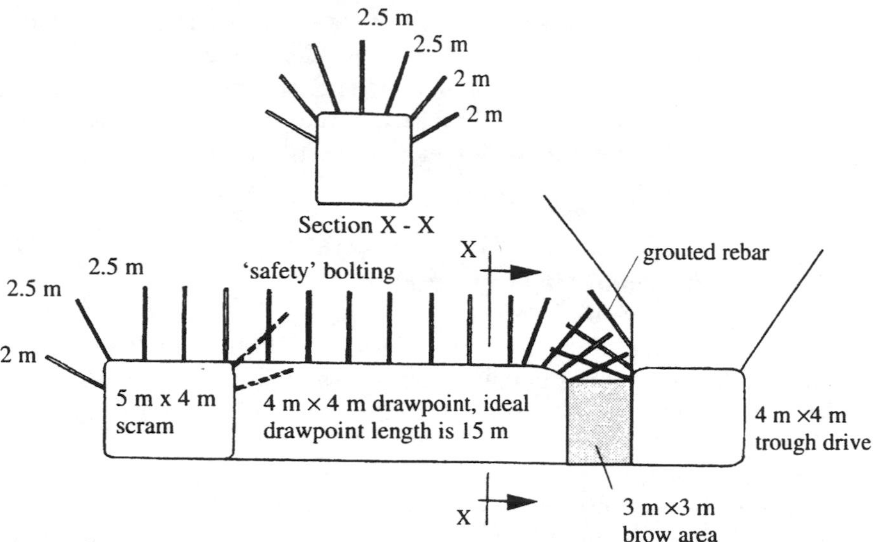

Figure 11.8: Suggested reinforcement for a drawpoint in a large mechanised mine. The brow area, shown shaded, is blasted last after the rebar has been grouted in place from the drawpoint and trough drive. 'Safety' bolting can be used in the drawpoint and scram.

Identification of weak zones in the rock and the provision of adequate reinforcement during construction are key elements in successful orepass design (Clegg and Hanson, 1992). Support, which will retain the rock close to the orepass surface without obstructing the passage of the ore, is required. Where possible, this should be installed from inside the orepass during excavation. Untensioned, fully grouted birdcage cables are probably the best type of reinforcement, since they have a high load carrying capacity for their whole length and the projecting ends will not obstruct the passage of the ore.

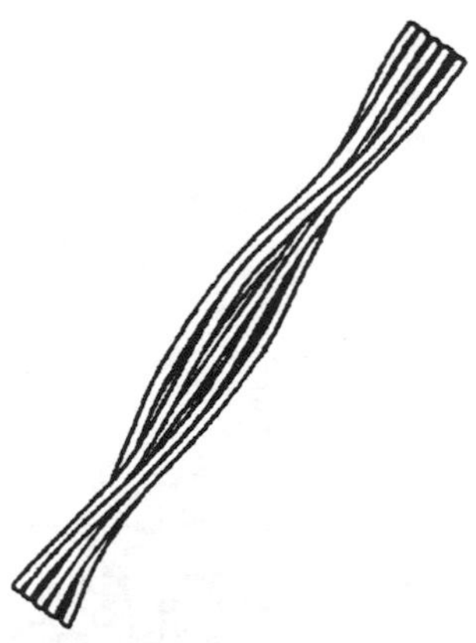

Birdcage cables are flexible but have a high load carrying capacity because the grout can penetrate into the cage

In many cases access may not be available to the inside of an orepass. The design of reinforcement is much more difficult, since there are generally only a few nearby openings from which reinforcement can be installed. Where an evaluation of the rock mass quality suggests that significant instability of the orepass walls may be a problem, the mining of special access drifts, from which reinforcement can be installed, may be required. While the cost of such excavations may be difficult to justify, experience has shown that the cost of orepass rehabilitation can be very high so that it is generally consider-

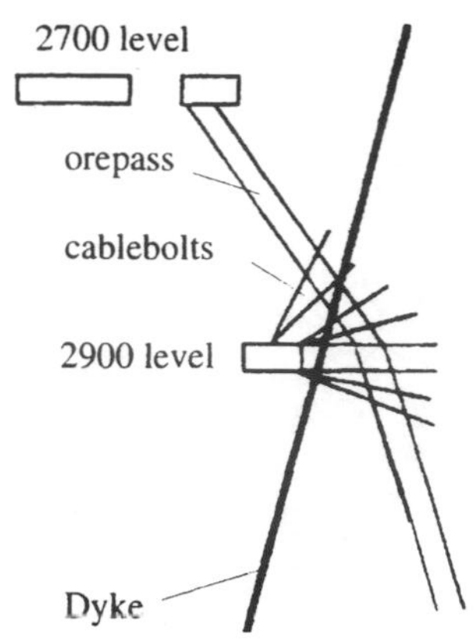

Pre-reinforcement of an orepass with cables installed from an external access. After Clegg and Hanson (1992).

ably cheaper to anticipate the problems and to provide pre-reinforcement for the surrounding rock mass.

The example illustrated in the margin sketch is from the paper by Clegg and Hanson, in which an orepass extension in granite in the Lockerby mine near Sudbury is described. Evaluation of the rock mass characteristics, based on rock mass classifications carried out on diamond drill core and exposures in adjacent openings, provided the basis for this support design. Because of high in situ stress conditions and anticipated slip on the dyke, it was decided to pre-reinforce the rock mass surrounding the orepass.

Birdcage cables, 12 m to 20 m long, were grouted into 60 mm diameter holes in order to provide support for the orepass/dyke intersection and for the orepass/level intersections. The cables for the orepass/dyke intersection were installed in fans spaced at 2.4 m, while the level intersections were supported by 12 m long cables on a 1.2 m × 1.2 m grid. Coated 'Swellex' bolts, 1.8 m long on a 1.2 m staggered pattern, were installed from inside the straight sections of the orepass, while resin grouted rebars were installed at the orepass bends.

11.5 Small openings in blocky rock

In many mining situations it is necessary to drive small openings parallel to the strike of dominant weakness planes in relatively massive rock. Two examples of such openings are illustrated in Figures 11.9 and 11.10.

Identification of potential wedges or blocks, which can slide or fall from the boundary of the opening, is an important first step in the design of reinforcement for this type of problem. The programs DIPS and UNWEDGE, described in previous chapters, were designed specifically for this type of problem and can be used to determine the size of wedges and the required support capacity.

For most mine openings of this type, ungrouted mechanically anchored bolts would be the obvious choice for support. Such bolts are simple and quick to install and can be tensioned to generate a positive clamping force on the potentially unstable wedge. This tension is important, since very little movement is required to separate the wedge from the surrounding rock. Once this happens, there is a potential for further loosening of the surrounding rock mass. Obviously, it is necessary to install these bolts before the entire perimeter of the wedge is exposed. This means that the bolts must be installed very close to the face as the drive is advanced.

When the opening is intended for long term use or where there is a risk of rapid corrosion due to the presence of acid mine water, the bolts should be fully grouted after tensioning. Tubular rockbolts, such as those manufactured by Stelco of Canada (see margin sketch) or Williams 'hollow-core' bolts, allow for simple grout injection where required.

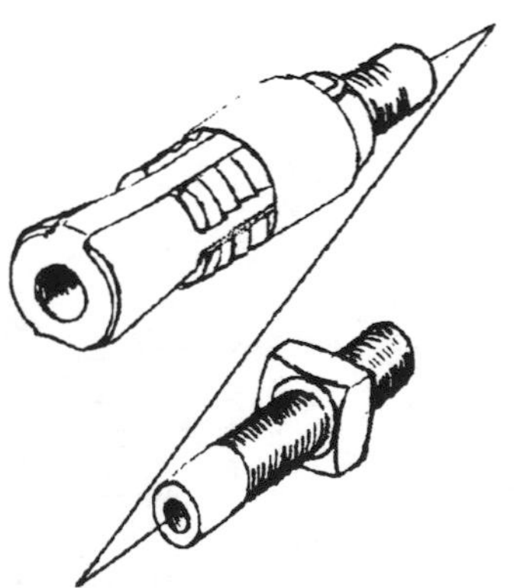

Stelpipe tubular rockbolt, manufactured by Stelco of Canada. The hole through the bolt simplifies the grouting process.

Figure 11.9: Profile of a small drive in massive blocky rock. The creation of a good excavation shape is sometimes difficult because blast fractures will tend to follow pre-existing weakness planes rather than break fresh rock.

Figure 11.10: Structurally controlled failures in an old slate quarry in Wales. No support was installed in this tunnel and the final profile is defined by the wedges which have fallen from the surrounding rock mass.

11.6 Small openings in heavily jointed rock

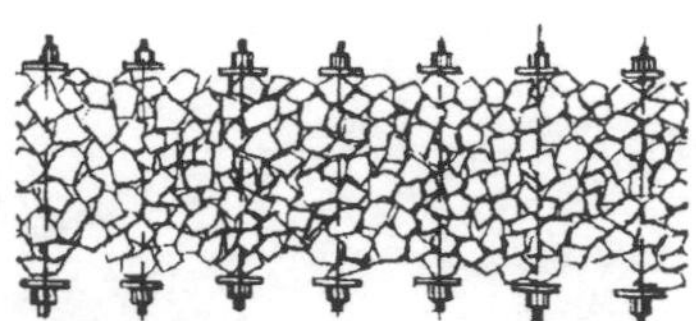

Section through Lang's bolted gravel plate model.

When the rock mass is heavily jointed and when failure can occur by ravelling (gravity driven) or squeezing (stress driven), support will only be effective if a pattern of rockbolts or dowels is used.

Lang (1961) demonstrated the effectiveness of pattern bolting by means of a model in which he created a self-supporting plate of bolted gravel. A simplified section through this model is given in the margin sketch and a photograph of the underside of the model is shown opposite. The model measured about 1.2 m × 1.2 m with a depth of gravel of about 200 mm. Miniature rockbolts, 7 mm in diameter and spaced on a 100 mm × 100 mm grid were used to bolt clean, angular gravel with an average grain size of about 30 mm. This model was not only self-supporting but proved capable of carrying a substantial load.

Photograph of the underside of Lang's bolted gravel plate model showing ravelling of rock pieces between face plates.

Based upon these model tests, Lang proposed that effective pattern bolting required the bolt length L to be at least twice the bolt spacing S and that the bolt spacing should not exceed about $4 \times B$, where B is the average size of potentially unstable blocks (see lower margin figure).

The principle of pattern support is illustrated in Figure 11.11 which shows a 4 m × 4 m excavation with 3 m long bolts on a 1.5 m × 1.5 m grid in the roof and upper sidewalls. The shaded zone in the figure represents compressive stress which is generated by tension in the bolts. Within this zone, the individual pieces of rock will remain interlocked and create a self-supporting arch. Note that the small triangular zones between the faceplates are not stressed and will tend to fall out unless supported by mesh or shotcrete.

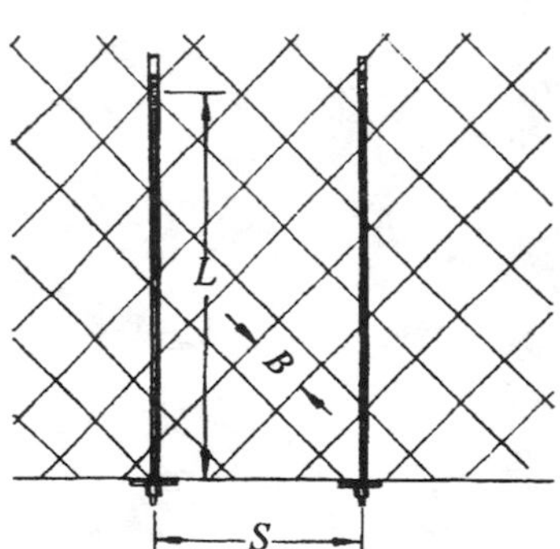

The effectiveness of pattern support is further demonstrated in Figure 11.12 which shows massive deformation in the roof and sidewalls of a rockbolted drive in heavily jointed rock in the Mount Isa mine in Australia.

In deciding upon the type of rockbolts to be used in these types of application, the stresses in the rock surrounding the excavation should be considered. When these stresses are low and when the primary failure process is one of gravity induced ravelling, light support in the form of ungrouted mechanically anchored bolts, grouted untensioned rebar or friction anchors such as Swellex or Split Set stabilisers will provide adequate support. When the stresses in the surrounding rock mass are high enough to induce squeezing and floor heave, heavier reinforcement is required. Grouted and tensioned rockbolts, fully grouted rebar with face plates or high axial load capacity Swellex bolts should be considered.

The program PHASES has been designed for the evaluation of support options in heavily jointed rock masses. An application of this program is illustrated in the following example.

A 5.5 m span tunnel with a slightly arched roof is excavated through a shear zone described as blocky/seamy rock of poor quality. The properties are defined by the following parameters:

Uniaxial compressive strength of intact rock	$\sigma_c = 70$ MPa
Hoek Brown constant	$m_b = 0.43$
Hoek-Brown constant	$a = 0.5$
Hoek-Brown constant	$s = 0$

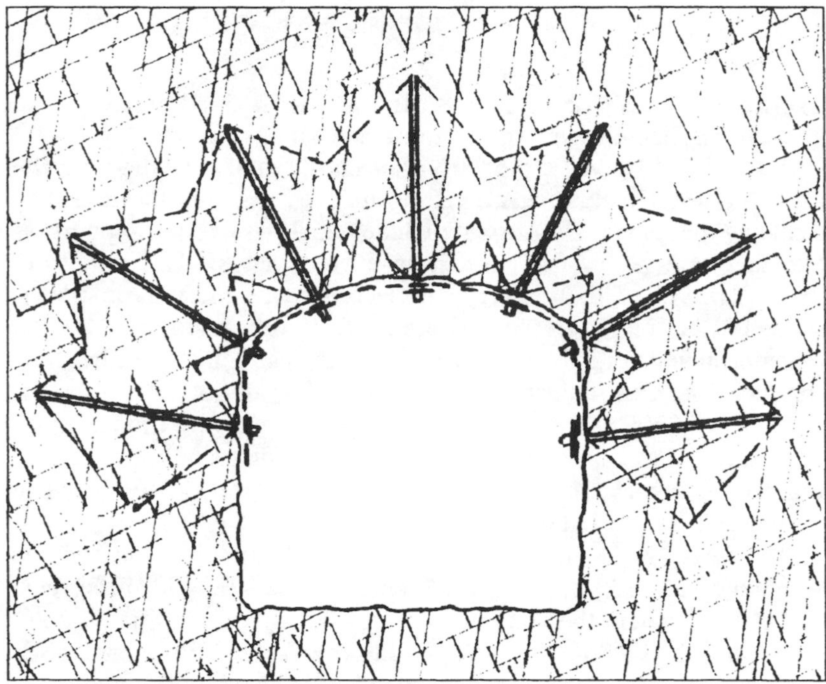

Figure 11.11: Pattern bolting for the support of heavily jointed rock which can fail by ravelling or squeezing. The shading represents a zone of compressive stress in which interlocking of individual rock pieces is retained and a self-supporting arch is created. Mesh or shotcrete should be applied to the excavation surface to retain small blocks and wedges in the stress-free zones between the rockbolts.

Figure 11.12: Roof deformation and sidewall failure in a drive in the 500 copper orebody in the Mount Isa mine in Australia. Closure between the roof and floor is more than 2 m but the rockbolts and weldmesh have prevented total collapse. After Mathews and Edwards (1969). Mount Isa Mines photograph.

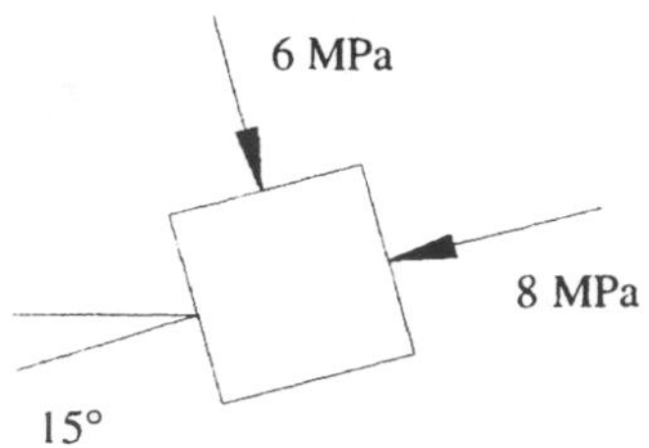

Modulus of deformation	E = 3000 MPa
Poisson's ratio	ν = 0.3

The maximum in situ stress is 8 MPa and is inclined at 15° to the horizontal. The minimum in situ stress is 6 MPa.

Figure 11.13 gives a plot, from an elastic analysis, showing the contours of available rock strength to induced stress in the rock surrounding the tunnel. As a first step in any analysis of rock support interaction, it is recommended that an elastic analysis be carried out using PHASES. This is a very simple procedure since the model is set up for the full progressive failure analysis, described later, but the material surrounding the excavation is defined as 'elastic' rather than 'plastic'. The elastic analysis takes only a few minutes to complete and it provides a useful check on the operation of the model. Once this analysis has been completed, the material surrounding the opening can be toggled to 'plastic' and, if required, the full progressive failure analysis carried out.

In Figure 11.13, the contour marked '1' encloses the rock in which the induced stresses exceed the available strength of the rock. In this case, where the zone of overstressed rock is significant compared with the size of the tunnel, a full progressive failure analysis is justified. When no overstressed zone appears or when the overstress is confined to small zones at the corners of the excavation, very little additional information will be gained from such an analysis.

Figure 11.14 gives the results of a PHASES analysis in which the rock surrounding the opening was defined as elastic-perfectly plastic. In other words, no brittle failure component was included in the analysis. In poor quality rock, such as that under consideration in this example, this assumption is justified since strength drop, after failure, is usually fairly small. This is in direct contrast to the very large strength drop associated with the failure of massive brittle rock, discussed in the previous chapter.

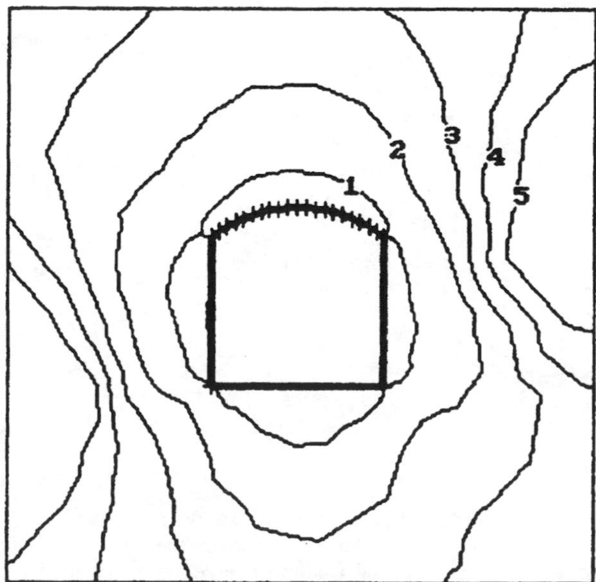

Figure 11.13: Contours of available strength to induced stress in the elastic rock surrounding a tunnel.

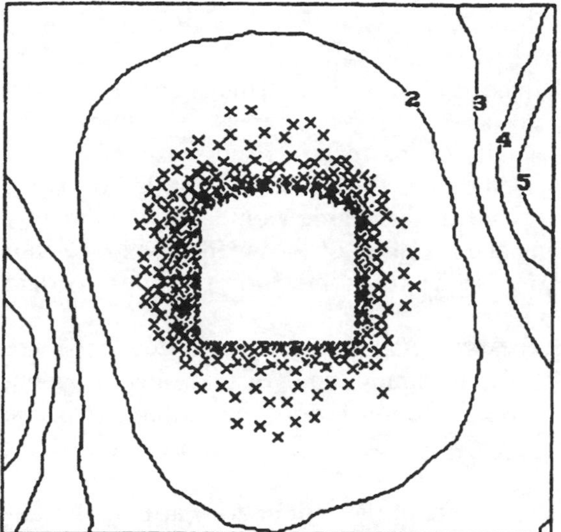

a) Failure zone, indicated by × symbols, and contours of the ratio of strength/stress in the rock mass surrounding the tunnel.

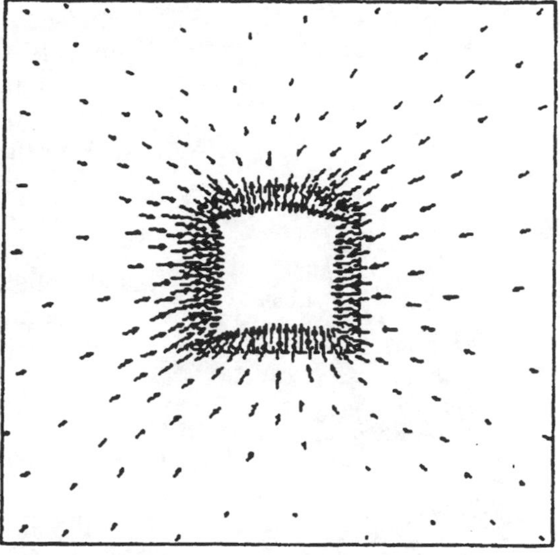

c) Displacements in the rock mass surrounding the unsupported tunnel. The maximum displacement in the roof is about 16 mm while the maximum floor heave is approximately 20 mm.

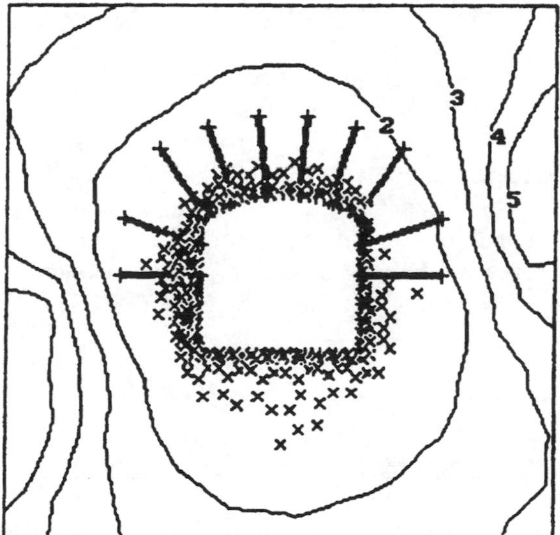

b) Failure zone and contours of strength/stress for the rock mass surrounding the rockbolted tunnel.

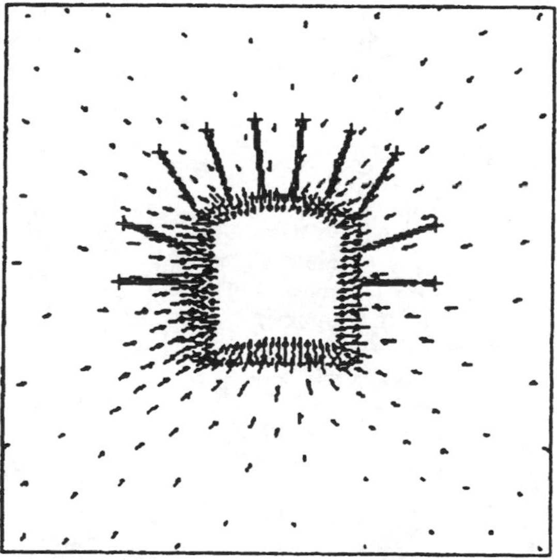

d) Displacements in the rock mass surrounding the rockbolted tunnel. The maximum displacement in the roof is about 11 mm while the maximum floor heave is approximately 17 mm.

Figure 11.14: Analysis of the influence of rockbolting the roof and upper sidewalls of a tunnel mined through poor quality blocky/seamy rock, typical of that which may be encountered in a fault or shear zone.

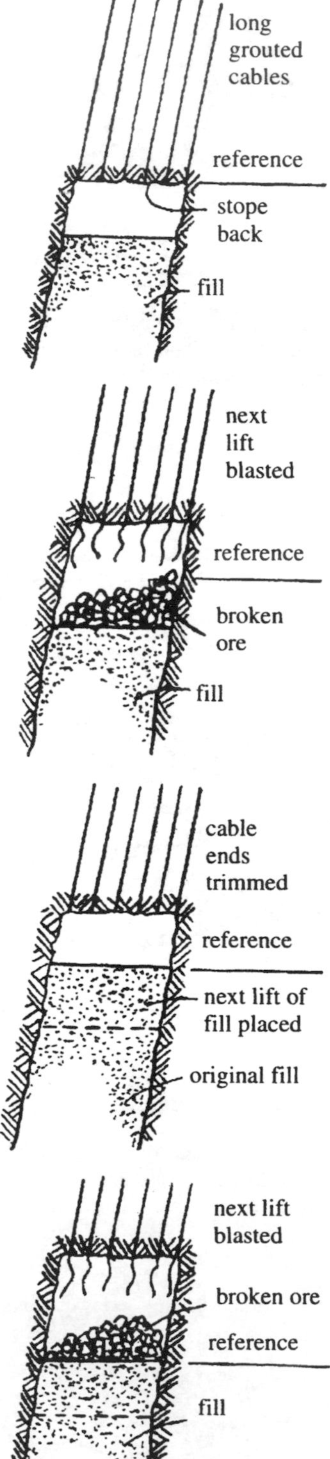

The failure zone in the rock mass surrounding the unsupported tunnel is shown in Figure 11.14a. Note that this zone extends beyond the overstressed zone defined in Figure 11.13. This is because the redistribution of stress, associated with the progressive failure of the rock in the immediate vicinity of the tunnel, results in a growth of the overstress zone indicated by the elastic analysis. A failure zone, as extensive as that indicated in this case, would almost certainly result in slabbing, spalling and ravelling of loosened rock in the roof and walls of the tunnel. Consequently, some form of support would be required for this tunnel.

Displacements in the rock surrounding the unsupported tunnel are shown in Figure 11.14c. The maximum roof displacement is about 16 mm while the floor heave is about 20 mm. This difference is due to the improvement in stress distribution resulting from the slight arching of the roof.

Figure 11.14b shows the extent of the failure zone after the installation of a pattern of 2.5 m long 25 mm diameter, mechanically anchored rockbolts placed on a 1 m × 1 m grid. These bolts have been installed in the roof and upper sidewalls of the tunnel and the reduction of the extent of the plastic zone is evident. Figure 11.14d gives the displacements for the supported tunnel and it will be noted that the maximum roof displacement has been reduced from 16 to 11 mm.

The rockbolt solution presented in Figure 11.14, with the addition of mesh, is probably adequate for most mining applications. However, in poor quality rock masses of the type considered in this example, mechanically anchored rockbolts may be ineffective because of slip of the anchors. In such cases, mechanical anchors can be replaced by anchors formed by inserting a fast-setting resin cartridge at the end of the hole. Alternatively, fully grouted, untensioned rebar can provide very effective support, provided that it is installed close to the advancing face. When floor heave is a problem, for example in the case of a tunnel for rail transportation where the stability of the floor is important, floor bolting or the casting of a concrete floor slab can be used to control failure of the rock in the floor.

11.7 Pre-support of openings

In the stopes used to extract the ore in an underground hard rock mine, safety of personnel and equipment and dilution of the ore due to failure of the surrounding rock mass are all major concerns. In room and pillar or cut and fill mining, in which personnel and equipment work in the stopes on a regular basis, safety is generally the primary objective. When non-entry bulk mining methods are used, dilution is the most critical factor when considering the stability of the rock mass surrounding the openings.

In small stopes, in which rockbolts and timber support have traditionally been used, the principles governing support design are similar to those already discussed in previous sections of this chapter. The discussion, which follows, deals mainly with stopes in which large volumes of rock can be involved in failure and where rockbolts and other 'light' support systems are not adequate. In most of these cases, cable bolting or backfilling are the principal support methods which the mine design engineer has available to control instability.

11.7.1 *Cut and fill stope support*

The essence of good stope support is to control the rock in the back and hanging wall before it is blasted and allowed to dilate and unravel. An illustration of this concept of pre-support is given in the margin sketch on the previous page which shows a cut and fill sequence in which long grouted cables are used to support the stope back. This procedure was adopted in Australia and Canada in the early 1970s and it involves grouting 15 to 20 m long untensioned cables into up-holes in the ore and/or hanging wall. As each successive cut is taken, the blasted ore strips off the ends of the cables but the remaining embedded lengths react to downward displacement of the rock mass and provide effective support for the back. The exposed cable ends are trimmed and, in some cases, intermediate rockbolts are placed to provide additional support. When several lifts have been taken and only 2 or 3 metres of cable remain in the back, a new set of overlapping cables is grouted in place before mining proceeds.

While this system of pre-reinforcement is very effective in rock masses of reasonable quality, the lack of face plates on the ends of the cables can cause problems in closely jointed rock. Ravelling of the unsupported rock in the immediate stope back is a safety hazard. Control of this loose rock requires expensive and time consuming bolting and meshing. A number of solutions to this problem have been tried, including the use of threaded bar in place of cables and the use of a variety of barrel and wedge attachments, which allow face plates to be attached to the ends of the cables.

A conventional barrel and wedge cable clamping device is illustrated in the margin sketch. These devices, which are manufactured and distributed by a number of companies, allow the cable to be tensioned after installation and before grouting.

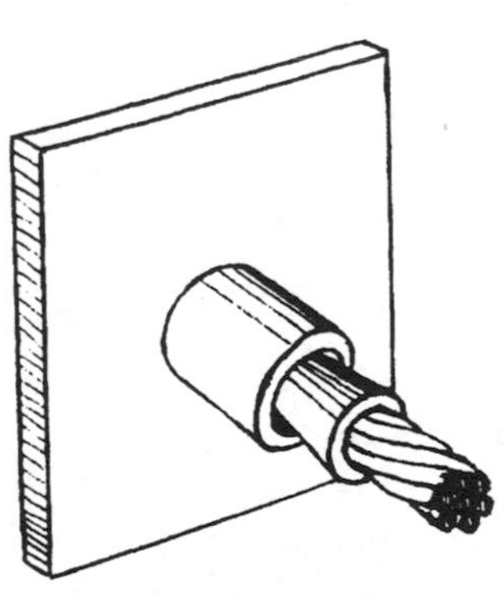

One of the simplest face plate attachments consists of a plate with a slotted hole into which a wedge is hammered. Any tendency for the rock to move down the cable forces the wedge further into the slot and tightens the grip on the cable end. While this system will not provide as high a load carrying capacity as the barrel and wedge, it is inexpensive and can be manufactured on most mine sites.

In general, tensioning of pre-placed cables is of little value and the downward movement of the rock mass, as mining proceeds, is sufficient to load the cables. In some cases a load of a few tonnes is applied to the cables to ensure that they are straight before they are grouted in place.

A large cut and fill stope in Mount Isa mine in Australia is illustrated in Figure 11.15. In this case, rockbolts have been used in both the back and the hanging wall to provide additional support. The density of this bolting is varied to suit local rock conditions and pre-installed cables have been successfully used in many of these stopes.

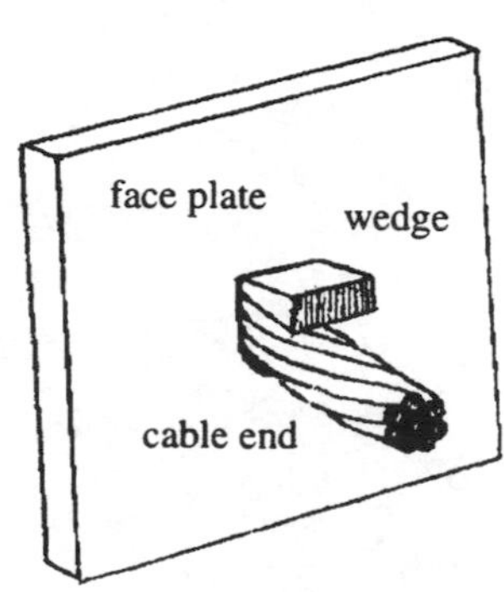

The control of roof failure by replacing conventional short rockbolts with long cablebolts is illustrated in Figures 11.16 and 11.17. The 'bench cut and fill' technique was employed in the Kotalahti mine in Finland (Lappalainen et al., 1984). The orebody was extracted from between layers of weak black schist by benching between two parallel drifts at 10 m vertical spacing. The drifts were bolted, using 2.4 m long rockbolts and, then shotcreted. This support system proved to be inadequate and was replaced by long cablebolts as illustrated in Figure 11.17. Although some cracking still occurred, the cablebolts prevented any further failure of the hangingwall.

Figure 11.15: Cut and fill stope in the Mount Isa mine, Australia. (Mount Isa Mine photograph).

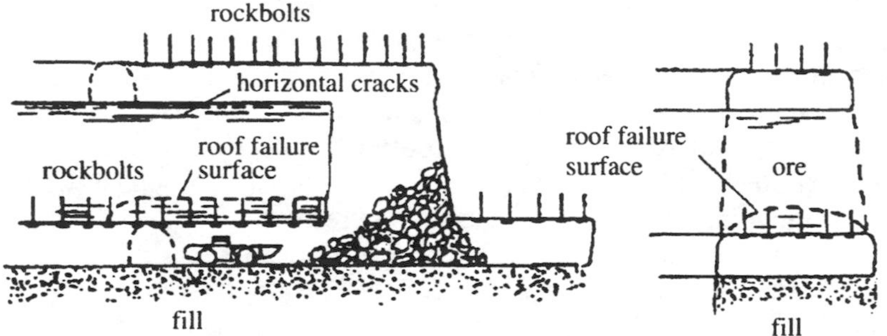

Figure 11.16: Roof failure in a bench cut and fill stope in the Kotalahti mine in Finland where only short rockbolts were used. After Lappalainen et al. (1984).

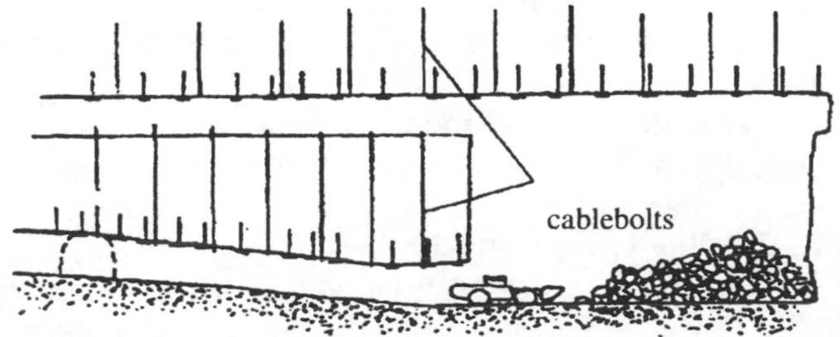

Figure 11.17: Use of cablebolts to control roof failure in bench cut and fill stope in the Kotalahti mine in Finland. After Lappalainen et al. (1984).

Pre-placed cablebolts, installed by means of an automated cablebolting machine, were used in the cut and fill mining of the Zirovski Vrh uranium mine in Slovenia (Bajzelj et al., 1992). Figure 11.18 shows the pattern of double cables (2 × 15.2 mm diameter cables) which were placed through the orebody. A high viscosity grout (water/cement ratio 0.3) was pumped into the holes using a two-stage pumping system. The grout tube, initially inserted to the end of the hole, was withdrawn as the hole was filled and the cables were then inserted through the grout.

Pull-out tests on the cablebolts, which had been instrumented with strain-gauges, were carried out. Strain-gauged cables were also monitored during blasting to determine the increase in axial load in the cables induced by the blasts. A finite element analysis of the mining sequence and support response, using non-linear stress-strain relationships for the rock mass, the fault zones and the backfill, was used to confirm the adequacy of the design. The authors of the paper explain that measurements, field testing and stress analysis were justified, since this mining and support technique were new to the Zirovski Vrh uranium mine and the aim of these studies was to optimise the mining method.

Bourchier et al. (1992) describe the use of 15 m long single 15.2 mm diameter seven strand cables for the support of cut and fill stopes in the Campbell mine near Balmertown in north-western Ontario, Canada. The placement of these cables is illustrated in Figure 11.19. The cablebolt spacings vary from 1.8 m × 2.4 m to 2.4 m × 2.4 m, depending upon the joint spacing, joint orientation and overall ground conditions. Initial support for the drift back is provided by means of 2.4 m long, mechanically anchored rockbolts on a 1.2 m × 1.2 m grid with weldmesh screen. The cables provide effective support for three 2.4 m lifts after which new cables are installed between the remainder of the previous cables. In some cases, the cables are recessed 2.4 m so that an additional lift can be mined before a new set of cables needs to be installed.

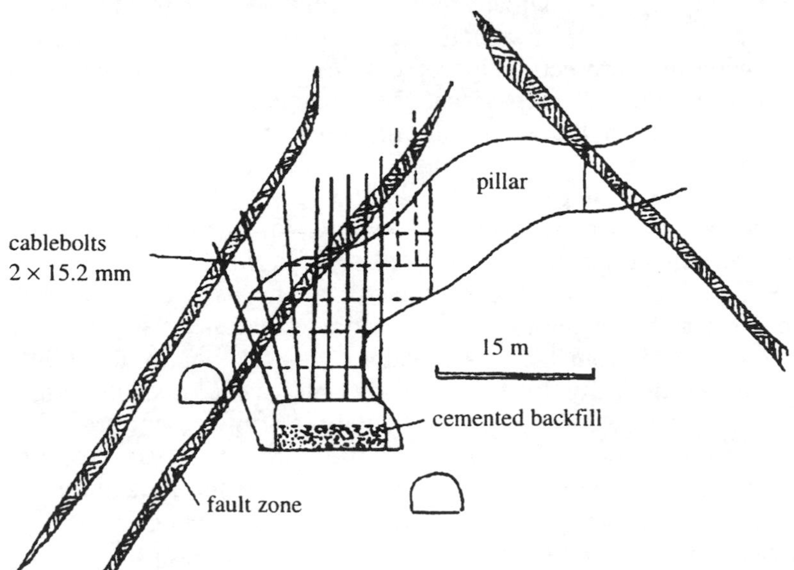

Figure 11.18: Placing of cables for cut and fill mining in the Zirovski Vrh uranium mine in Slovenia. After Bajzelj et al. (1992).

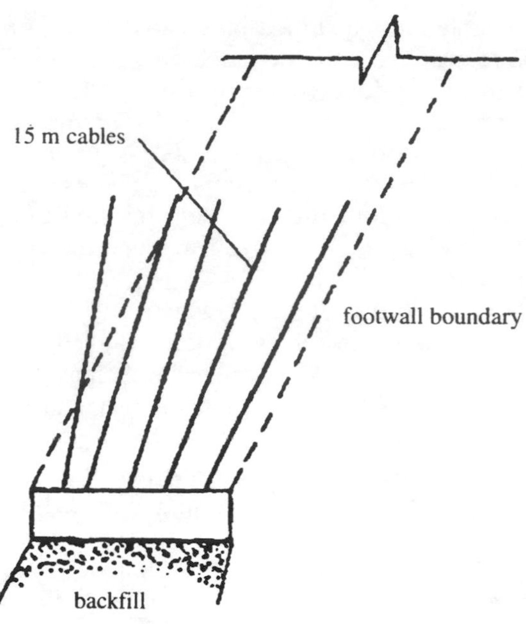

Figure 11.19: Placing of cables to provide support for both the orebody and the hangingwall in cut and fill stopes at the Campbell mine. After Bourchier et al. (1992).

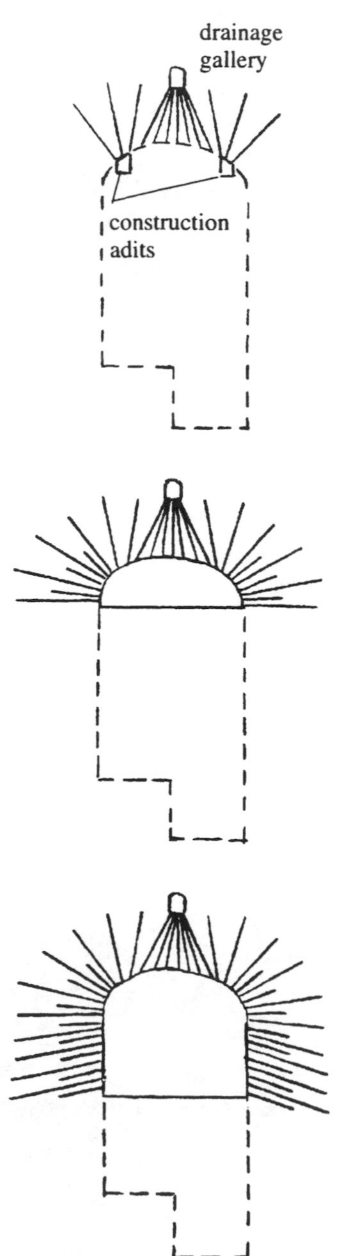

The cables are untensioned and fully grouted and Bourchier at al state that experience has shown that this system provides rock mass reinforcement superior to other support systems which have been tried.

11.7.2 *Pre-reinforcement of permanent openings*

Pre-reinforcement is not restricted to cut and fill mining. It has been successfully applied in many other mining and civil engineering projects. For example, in mining a large 'permanent' excavation for an underground crusher station or garage, it may be appropriate to pre-reinforce the rock mass around the opening.

An example of the application of pre-reinforcement on a large civil engineering project is illustrated in the series of margin sketches opposite. This example is based on the construction of a 22 m span × 45 m high power cavern in bedded sandstone for the Mingtan hydro-electric project in Taiwan (Hoek and Moy, 1993).

Before the main construction contract commenced, the rock mass above the arch was reinforced from a drainage gallery 10 m above the crown and from two construction adits, shown in the upper margin sketch. Fifty tonne capacity cables were installed on a 2 m × 2 m grid pattern and a straightening load of 5 tonnes was applied to each cable before grouting. The purpose of this pre-reinforcement was to improve the overall quality of the rock mass, so that the main contract could proceed without the delays caused by the need to support unstable areas in the immediate roof rock.

Once the main contract commenced, the roof arch was opened to full span as shown in the centre margin sketch. As each cable end was exposed in the centre of the arch, faceplates were attached by means of barrel and wedge anchors. A load of 10 tonnes was applied during the faceplate installation to ensure a positive anchorage. The projecting cable ends were then trimmed and a layer of 50 mm thick steel fibre reinforced shotcrete was applied. Where required, 5 m long 25 mm diameter mechanically anchored and grouted rockbolts were installed between the cables.

After completion of the roof arch, the remainder of the cavern was excavated using 2.5 m vertical benches. Twelve to fifteen metre long, 112 tonne capacity, corrosion protected cables were installed at a downward inclination of 15° on a 3 m × 3 m grid in the sidewalls. Before grouting, these cables were tensioned to an average of 40% of

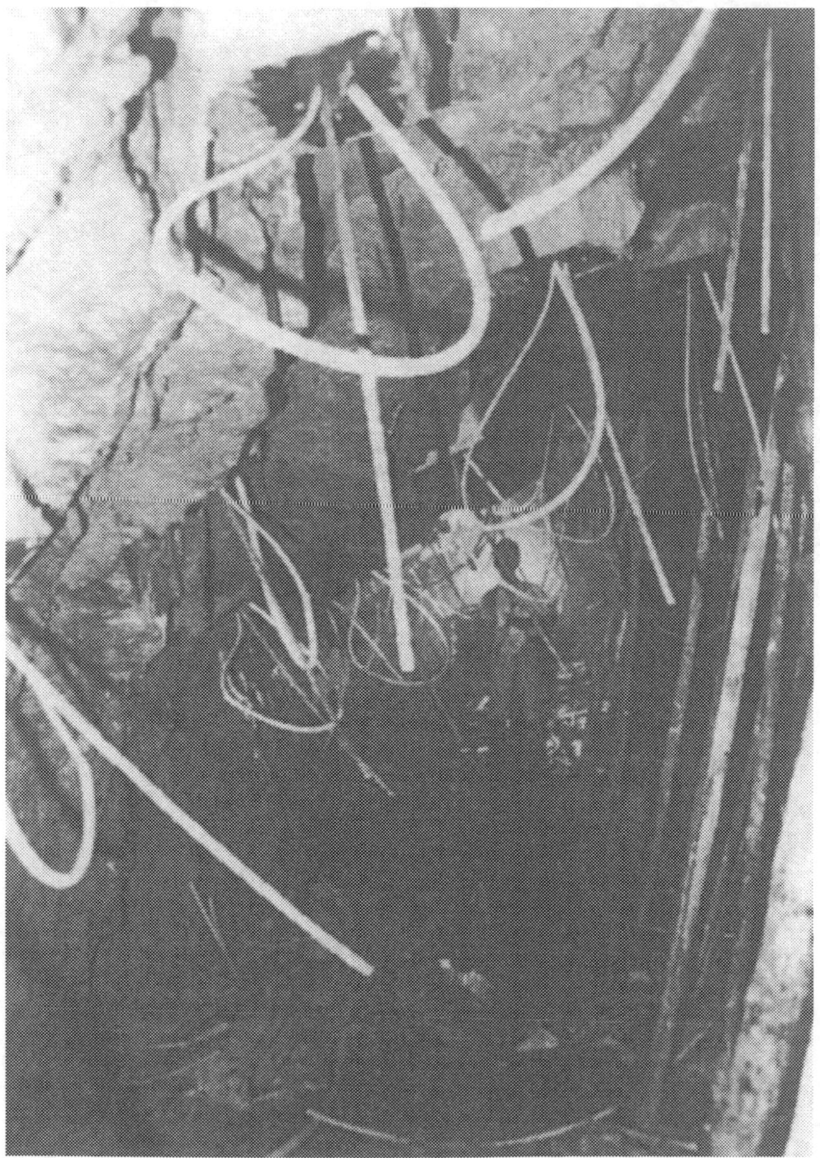

Figure 11.20: Cables installed from one of the construction galleries in the Mingtan project in Taiwan. A straightening load of 5 tonnes was applied before these cables were grouted. Face plates were installed on all exposed cable ends.

their yield strength. Intermediate rockbolts, 6 m long 25 mm diameter, were installed and tensioned before grouting. Finally, a steel fibre reinforced shotcrete layer of 50 mm thickness was applied to the sidewalls. The shotcrete on the upper sidewalls and roof arch was built up to a maximum of 150 mm thickness.

This reinforcing system proved to be very effective in controlling the extent of failure and the deformations in the rock mass surrounding the cavern. A maximum displacement of 78 mm was recorded in the sidewalls of the cavern and very little additional deformation has occurred since completion of construction.

Figure 11.20 shows the cables installed from the construction adits for the pre-reinforcement of the rock mass above the crown. Installation of the cables in the cavern sidewalls is illustrated in Figure 11.21.

Figure 11.21: Installation of cables in the sidewall of the power cavern of the Mingtan project in Taiwan.

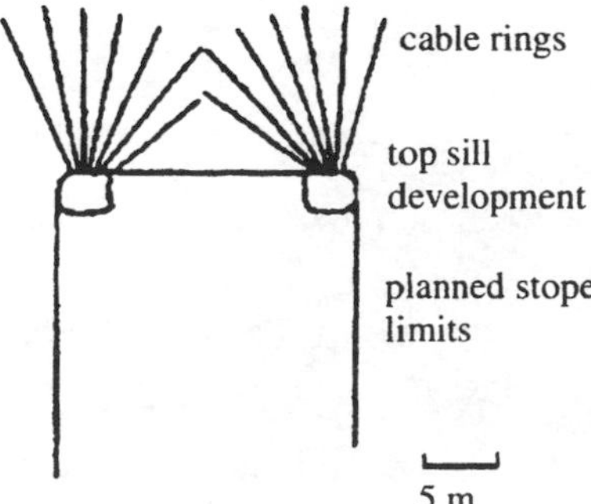

Figure 11.22: Typical cable fans in the back of a wide open stope. After Fuller (1984).

11.7.3 *Reinforcement of non-entry stopes*

The mining of large orebodies by means of non-entry stopes results in a significant reduction of the exposure of personnel and equipment to rockfalls and cave-ins. On the other hand, dilution of the ore from rock mass failure in the stope back and hanging wall can give rise to serious economic problems. The installation of support is one of the main tools available to the mining engineer for the control of this dilution.

Fuller (1984) points out that the placement of a uniform array of cables, such as those used in cut and fill mining, is seldom practical in open stoping because of access limitations. In the case of small open stopes, access from the top sill can be used to provide a reasonably uniform distribution of cables but, for wide open stopes, relatively wide spans remain unsupported as shown in Figure 11.22.

In the case of stope walls, access is even more difficult and dilution due to overbreak can be a serious problem. Figure 11.23 illustrates hanging wall overbreak in the J704 stope, at Mount Isa mine in Australia, where radiating cable rings from available hanging wall drifts were used to provide support (Bywater and Fuller, 1984). An alternative design, where these radial rings are supplemented by cable fans from specially driven access tunnels, is illustrated in Figure 11.24. Clearly, the cost of providing special access is considerable but it can be justified if a significant reduction in overbreak can be achieved.

Figure 11.25 illustrates the installation of cables from a bottom sill and sublevel to achieve an even cable density in the hangingwall of an open stope. Dashed lines indicate sections of the holes in which the cables have been countersunk.

Bourchier et al. (1992) describe the use of a similar cable installations to support the hanging walls of longhole stopes in the Campbell mine. These installations are shown in Figure 11.26.

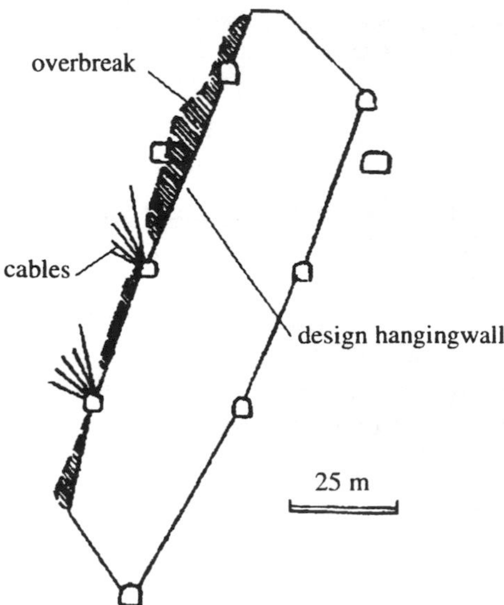

Figure 11.23: Hangingwall overbreak in the J704 stope at the Mount Isa mine. After Bywater and Fuller (1984).

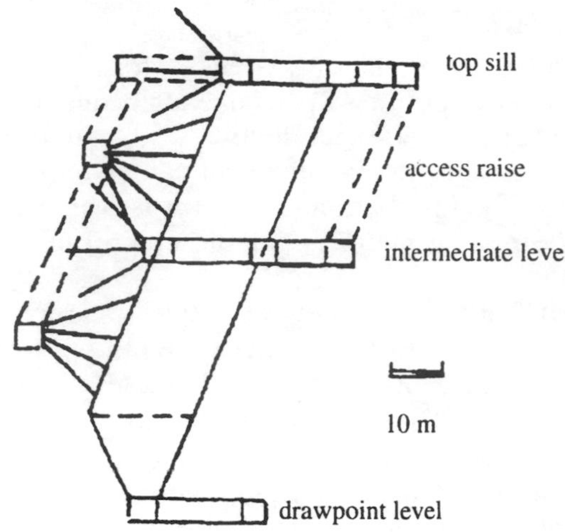

Figure 11.24: Cable patterns from specially driven access and from hangingwall drilling access. After Fuller (1984).

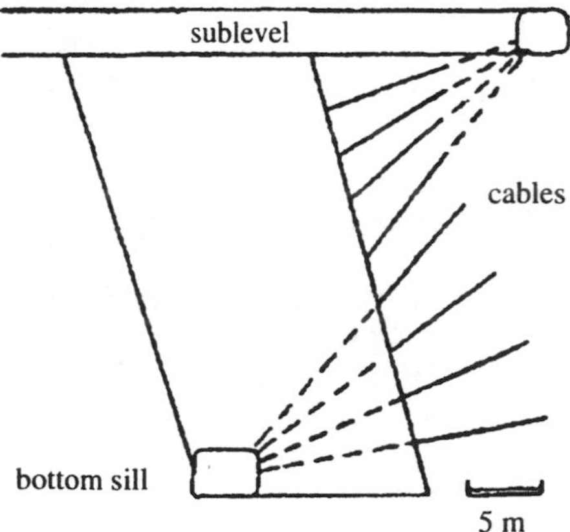

Figure 11.25: Fanning cables from a top and bottom access provides an even cable density in the hangingwall. After Fuller (1984).

Figure 11.27 shows a cable installation for sublevel stoping in the Kotalahti mine in Finland (Lappalainen et al., 1984). Two 15.2 mm diameter cables with a total capacity of 50 tonnes were fully grouted into each hole. Cables were not installed in holes through orebody sections, shown as dashed lines in the figure. A fully mechanised jumbo was used to drill the holes, install, cut and grout the cables. Cables of up to 50 m in length have been successfully installed with this machine.

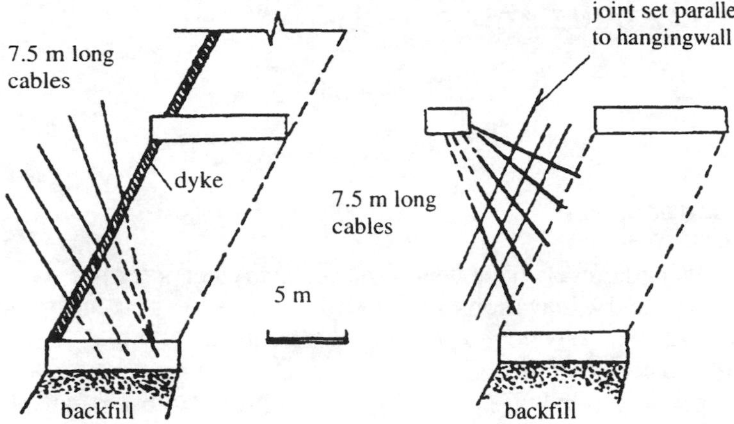

Figure 11.26: Hangingwall support for a longhole stope by means of cables placed from inside a stope and from a bypass drift at the Campbell mine. After Bourchier et al.(1992).

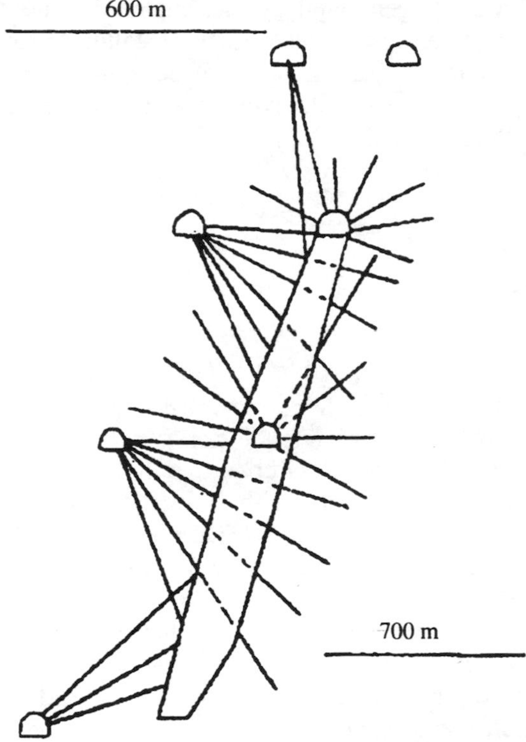

Figure 11.27: Cable bolt placing for sublevel stoping in the Kotalahti mine in Finland. Orebody sections are left without cables. After Lappalainen et al. (1984).

12 Rockbolts and dowels

12.1 Introduction

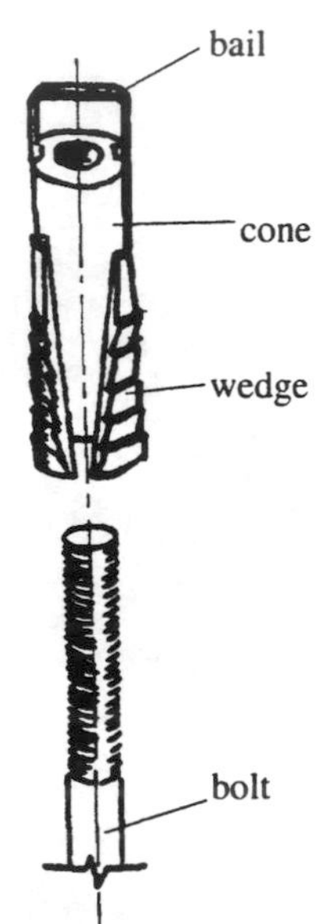

Rockbolts and dowels have been used for many years for the support of underground excavations and a wide variety of bolt and dowel types have been developed to meet different needs which arise in mining and civil engineering.

Rockbolts generally consist of plain steel rods with a mechanical anchor at one end and a face plate and nut at the other. They are always tensioned after installation. For short term applications the bolts are generally left ungrouted. For more permanent applications or in rock in which corrosive groundwater is present, the space between the bolt and the rock can be filled with cement or resin grout.

Dowels or anchor bars generally consist of deformed steel bars which are grouted into the rock. Tensioning is not possible and the load in the dowels is generated by movements in the rock mass. In order to be effective, dowels have to be installed before significant movement in the rock mass has taken place. Figure 12.1 illustrates a number of typical rockbolt and dowel applications which can be used to control different types of failure which occur in rock masses around underground openings.

12.2 Rockbolts

12.2.1 *Mechanically anchored rockbolts*

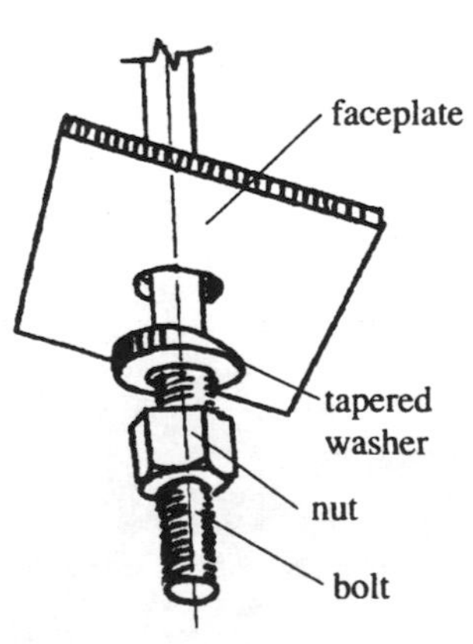

Expansion shell rockbolt anchors come in a wide variety of styles but the basic principle of operation is the same in all of these anchors. As shown in the margin sketch, the components of a typical expansion shell anchor are a tapered cone with an internal thread and a pair of wedges held in place by a bail. The cone is screwed onto the threaded end of the bolt and the entire assembly is inserted into the hole which has been drilled to receive the rockbolt. The length of the hole should be at least 100 mm longer than the bolt otherwise the bail will be dislodged by being forced against the end of the hole. Once the assembly is in place, a sharp pull on the end of the bolt will seat the anchor. Tightening the bolt will force the cone further into the wedge thereby increasing the anchor force. These expansion shell anchors work well in hard rock but they are not very effective in closely jointed rocks and in soft rocks, because of deformation and failure of the rock in contact with the wedge grips. In such rocks, the use of resin cartridge anchors, described later in this chapter, are recommended.

At the other end of the rockbolt from the anchor, a fixed head or threaded end and nut system can be used. In either case, some form of faceplate is required to distribute the load from the bolt onto the rock face. In addition, a tapered washer or conical seat is needed to

	Low stress levels	High stress levels
Massive rock	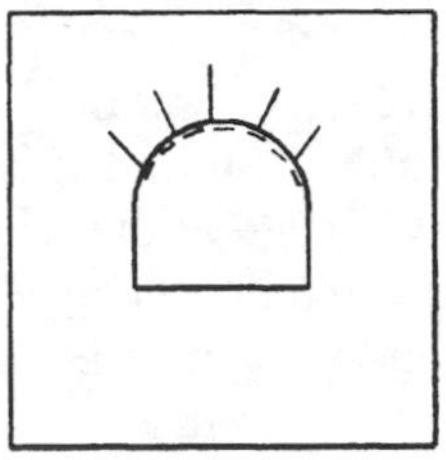 Massive rock subjected to low in situ stress levels. No support or 'safety bolts' or dowels and mesh.	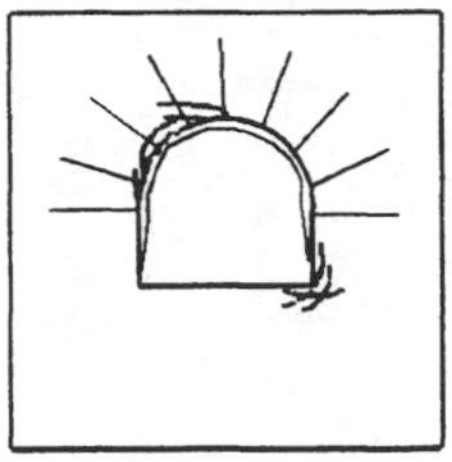 Massive rock subjected to high in situ stress levels. Pattern rockbolts or dowels with mesh or shotcrete to inhibit fracturing and to keep broken rock in place.
Jointed rock	 Massive rock with relatively few discontinuities subjected to low in situ stress conditions. 'Spot' bolts located to prevent failure of individual blocks and wedges. Bolts must be tensioned.	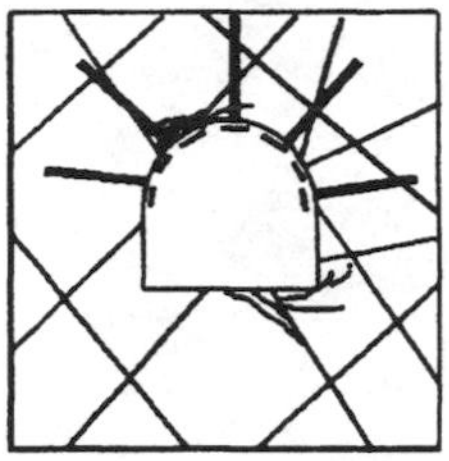 Massive rock with relatively few discontinuities subjected to high in situ stress conditions. Heavy bolts or dowels, inclined to cross rock structure, with mesh or steel fibre reinforced shotcrete on roof and sidewalls.
Heavily jointed rock	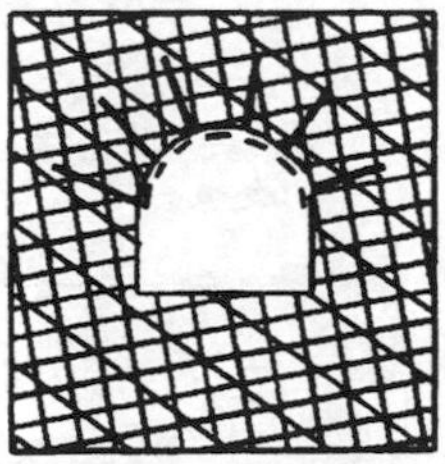 Heavily jointed rock subjected to low in situ stress conditions. Light patter bolts with mesh and/or shotcrete will control ravelling of near surface rock pieces.	 Heavily jointed rock subjected to high in situ stress conditions. Heavy rockbolt or dowel pattern with steel fibre reinforced shotcrete. In extreme cases, steel sets with sliding joints may be required. Invert struts or concrete floor slabs may be required to control floor heave.

Figure 12.1: Typical rockbolt and dowel applications to control different types of rock mass failure.

compensate for the fact that the rock face is very seldom at right angles to the bolt. A wide variety of faceplates and tapered or domed washers are available from rockbolt suppliers.

In general, threads on rockbolts should be as coarse as possible and should be rolled rather than cut. A fine thread is easily damaged and will cause installation problems in a typical mine environment. A cut thread weakens the bolt and it is not unusual to see bolts with cut threads which have failed at the first thread at the back of the nut. Unfortunately, rolled thread bolts are more expensive to manufacture and the added cost tends to limit their application to situations where high strength bolts are required.

Tensioning of rockbolts is important to ensure that all of the components are in contact and that a positive force is applied to the rock. In the case of light 'safety' bolts, the amount of tension applied is not critical and tightening the nut with a conventional wrench or with a pneumatic torque wrench is adequate. Where the bolts are required to carry a significant load, it is generally recommended that a tension of approximately 70% of the capacity of the bolt be installed initially. This provides a known load with a reserve in case of additional load being induced by displacements in the rock mass.

One of the primary causes of rockbolt failure is rusting or corrosion and this can be counteracted by filling the gap between the bolt and the drillhole wall with grout. While this is not required in many mining situations, grouting should be considered where the groundwater is likely to induce corrosion or where the bolts are required to perform a 'permanent' support function.

Figure 12.2: Use of a torque wrench to tension a rockbolt. Rockbolt manufacturers will supply torque-tension calibration curves on request. These calibrations differ, depending upon the thread type used on the bolt.

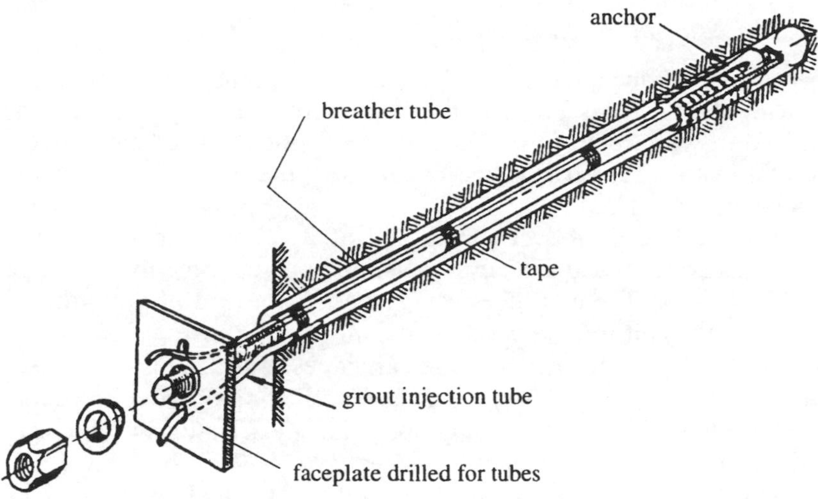

Figure 12.3: Grout injection arrangements for a mechanically anchored rockbolt.

The traditional method of grouting uphole rockbolts is to use a short grout tube to feed the grout into the hole and a smaller diameter breather tube, extending to the end of the hole, to bleed the air from the hole. The breather tube is generally taped to the bolt shank and this tends to cause problems because this tube and its attachments can be damaged during transportation or insertion into the hole. In addition, the faceplate has to be drilled to accommodate the two tubes, as illustrated in Figure 12.3. Sealing the system for grout injection can be a problem.

Many of these difficulties are overcome by using a hollow core bolt, as illustrated in the lower margin sketch on page 136. While more expensive than conventional bolts, these hollow bolts make the grouting process much more reliable and should be considered wherever permanent rockbolt installations are required. The grout should be injected through a short grout tube inserted into the collar of the hole and the central hole in the bolt should be used as a breather tube. When installing these bolts in downholes, the grout should be fed through the bolt to the end of the hole and the short tube used as a breather tube.

Since the primary purpose of grouting mechanically anchored bolts is to prevent corrosion and to lock the mechanical anchor in place, so that it cannot be disturbed by blasting vibrations and rock mass displacement, the strength requirement for the grout is not as important as it is in the case of grouted dowels or cables (to be discussed later). The grout should be readily pumpable without being too fluid and a typical water/cement ratio of 0.4 to 0.5 is a good starting point for a grout mix for this application. It is most important to ensure that the annular space between the bolt and the drillhole wall is completely filled with grout. Pumping should be continued until there is a clear indication that the air has stopped bleeding through the breather tube or that grout is seen to return through this tube.

12.2.2 *Resin anchored rockbolts*

Mechanically anchored rockbolts have a tendency to work loose when subjected to vibrations due to nearby blasting or when anchored in weak rock. Consequently, for applications where it is essential that the support load be maintained, the use of resin anchors should be considered.

A typical resin product is made up of two component cartridges containing a resin and a catalyst in separate compartments, as shown in Figure 12.4. The cartridges are pushed to the end of the drillhole ahead of the bolt rod which is then spun into the resin cartridges by the drill. The plastic sheath of the cartridges is broken and the resin and catalyst mixed by this spinning action. Setting of the resin occurs within a few minutes (depending upon the specifications of the resin mix) and a very strong anchor is created.

This type of anchor will work in most rocks, including the weak shales and mudstones in which expansion shell anchors are not suitable. For 'permanent' applications such as bolting around shaft stations or crusher chambers, consideration should be given to the use of fully resin-grouted rockbolts, illustrated in Figure 12.5. In these applications, a number of slow-setting resin cartridges are inserted into the drillhole behind the fast-setting anchor cartridges. Spinning the bolt rod through all of these cartridges initiates the chemical reaction in all of the resins but, because the slow-setting 'grout' cartridges are timed to set in up to 30 minutes, the bolt can be tensioned within two or three minutes of installation (after the fast anchor resin has set). This tension is then locked in by the later-setting grout cartridges and the resulting installation is a fully tensioned, fully grouted rockbolt.

The high unit cost of resin cartridges is offset by the speed of installation. The process described above results in a completely tensioned and grouted rockbolt installation in one operation, something that cannot be matched by any other system currently on the market. However, there are potential problems with resins.

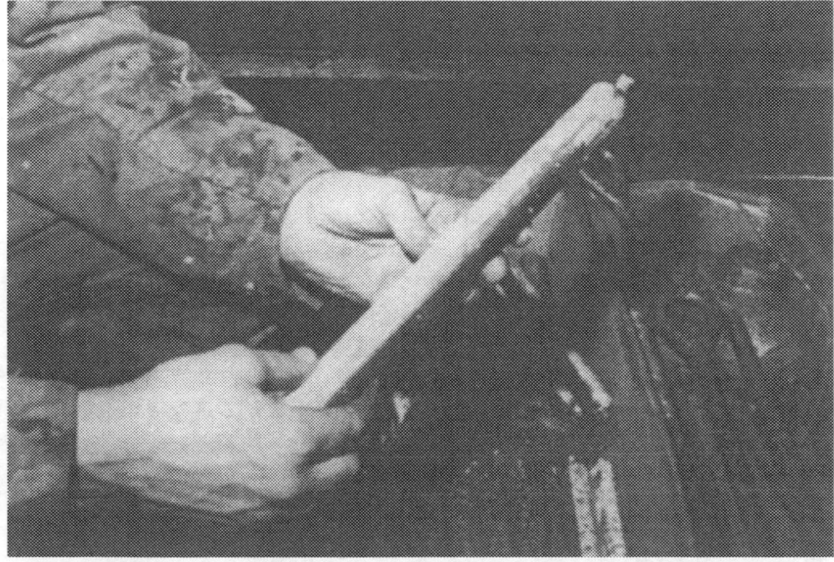

Figure 12.4: Typical resin cartridge for use in anchoring and grouting rockbolts.

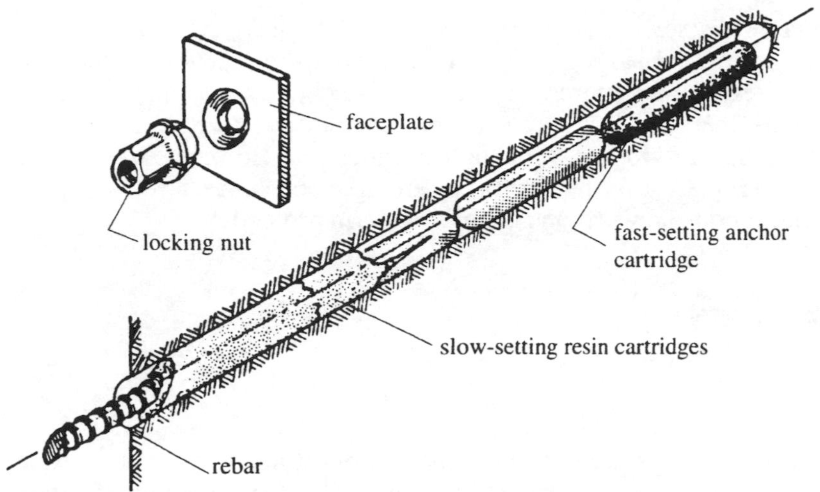

Figure 12.5: Typical set-up for creating a resin anchored and grouted rockbolt. Resin grouting involves placing slow-setting resin cartridges behind the fast-setting anchor cartridges and spinning the bolt rod through them all to mix the resin and catalyst. The bolt is tensioned after the fast-setting anchor resin has set and the slow-setting resin sets later to grout the rod in place.

Most resin/catalyst systems have a limited shelf life which, depending upon storage temperatures and conditions, may be as short as six months. Purchase of the resin cartridges should be limited to the quantities to be used within the shelf life. Care should be taken to store the boxes under conditions which conform to the manufacturer's recommendations. In critical applications, it is good practice to test the activity of the resin by sacrificing one cartridge from each box, before the contents are used underground. This can be done by breaking the compartment separating the resin and catalyst by hand and, after mixing the components, measuring the set time to check whether this is within the manufacturer's specifications.

Breaking the plastic sheath of the cartridges and mixing the resins effectively can also present practical problems. Cutting the end of the bolt rod at an angle to form a sharp tapered point will help in this process, but the user should also be prepared to do some experimentation to achieve the best results. Note that the length of time or the number of rotations for spinning the resins is limited. Once the setting process has been initiated, the structure of the resin can be damaged and the overall installation weakened by additional spinning. Most manufacturers supply instructions on the number of rotations or the length of time for spinning.

In some weak argillaceous rocks, the drillhole surfaces become clay-coated during drilling. This causes slipping of the resin cartridges during rotation, resulting in incomplete mixing and an unsatisfactory bond. In highly fractured rock masses, the resin may seep into the surrounding rock before setting, leaving voids in the resin column surrounding the rockbolt. In both of these cases, the use of cement grouting rather than resin grouting may provide a more effective solution.

There is some uncertainty about the long-term corrosion protection offered by resin grouts and also about the reaction of some of these resins with aggressive groundwater. For typical mining applications, these concerns are probably not an issue because of the limited design life for most rockbolt installations. However, where very long service life is required, current wisdom suggests that cement grouted bolts may provide better long term protection.

12.3 Dowels

12.3.1 *Grouted dowels*

When conditions are such that installation of support can be carried out very close to an advancing face, or in anticipation of stress changes which will occur at a later mining stage, dowels can be used in place of rockbolts. The essential difference between these systems is that tensioned rockbolts apply a positive force to the rock, while dowels depend upon movement in the rock to activate the reinforcing action. Drawpoints, which are mined before the overlying stopes are blasted, are good examples of excavations where untensioned grouted dowels will work well.

The simplest form of dowel in use today is the cement grouted dowel as illustrated in Figure 12.6. A thick grout (typically a 0.3 to 0.35 water/cement ratio grout) is pumped into the hole by inserting the grout tube to the end of the hole and slowly withdrawing the tube as the grout is pumped in. Provided that a sufficiently viscous grout is used, it will not run out of the hole. The dowel is pushed into the hole about half way and then given a slight bend before pushing it fully into the hole. This bend will serve to keep the dowel firmly lodged in the hole while the grout sets. Once the grout has set, a face plate and nut can be fitted onto the end of the dowel and pulled up tight. Placing this face place is important since, if the dowel is called on to react to displacements in the rock mass, the rock close to the borehole collar will tend to pull away from the dowel unless restrained by a faceplate.

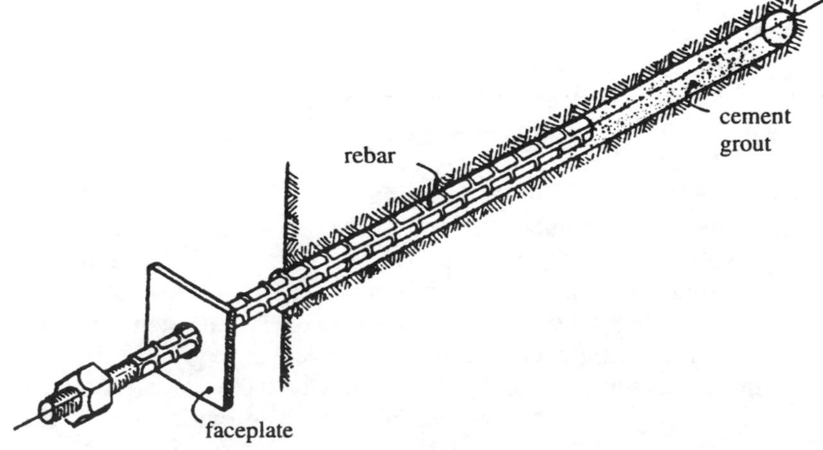

Figure 12.6: Grouted dowel using a deformed bar inserted into a grout-filled hole.

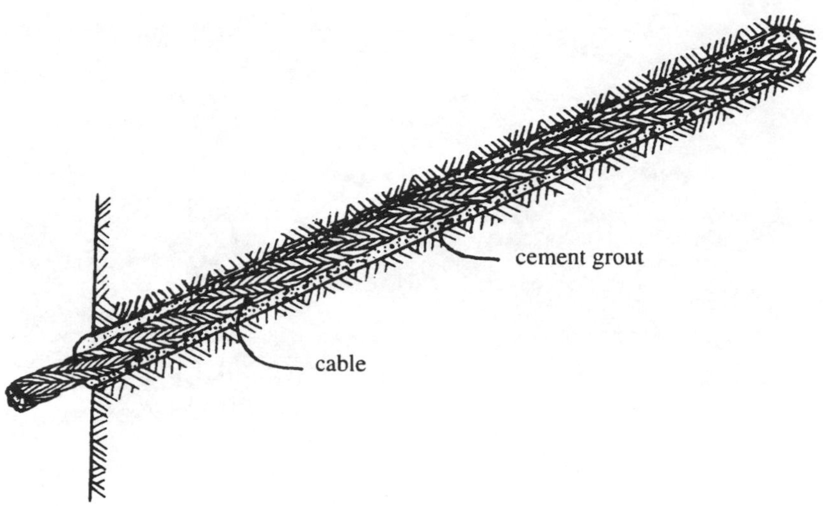

Figure 12.7: Grouted cables can be used in place of rebar when more flexible support is required or where impact and abrasion cause problems with rigid support.

In drawpoints and ore-passes, the flow of broken rock can cause serious abrasion and impact problems. The projecting ends of grouted rebars can obstruct the flow of the rock. Alternatively, the rebar can be bent, broken or ripped out of the rock mass. In such cases, grouted flexible cable, illustrated in Figure 12.7, can be used in place of the more rigid rebar. This will allow great flexibility with impact and abrasion resistance.

Older type grouted dowels such as the Scandinavian 'perfobolt' or dowels, where the grout is injected after the rod has been inserted, tend not to be used. The installation is more complex and time consuming and the end product does not perform any better than the simple grouted dowel described above.

12.3.2 *Friction dowels or 'Split Set' stabilisers*

Split Set stabilisers were originally developed by Scott (1976, 1983) and are manufactured and distributed by Ingersoll-Rand. The system, illustrated in Figure 12.8, consists of a slotted high strength steel tube and a face plate. It is installed by pushing it into a slightly undersized hole and the radial spring force generated, by the compression of the *C* shaped tube, provides the frictional anchorage along the entire length of the hole. A list of typical Split Set stabiliser dimensions and capacities is given in Table 12.1.

Because the system is very quick and simple to install, it has gained rapid acceptance by miners throughout the world. The device is particularly useful in mild rockburst environments, because it will slip rather than rupture and, when used with mesh, will retain the broken rock generated by a mild burst. Provided that the demand imposed on Split Sets stabilisers does not exceed their capacity, the system works well and can be considered for many mining applications.

Figure 12.8: Split Set stabiliser. Ingersoll-Rand photograph.

Table 12.1: Split Set specifications (After Split Set Division, Ingersoll-Rand Company).

Split Set stabiliser model	SS-33	SS-39	SS-46
Recommended nominal bit size	31 to 33 mm	35 to 38 mm	41 to 45 mm
Breaking capacity, average	10.9 tonnes	12.7 tonnes	16.3 tonnes
minimum	7.3 tonnes	9.1 tonnes	13.6 tonnes
Recommended initial anchorage (tonnes)	2.7 to 5.4	2.7 to 5.4	4.5 to 8..2
Tube lengths	0.9 to 2.4 m	0.9 to 3.0 m	0.9 to 3.6 m
Nominal outer diameter of tube	33 mm	39 mm	46 mm
Domed plate sizes	150 × 150 mm 125 × 125 mm	150 × 150 mm 125 × 125 mm	150 × 150 mm
Galvanised system available	yes	yes	yes
Stainless steel model available	no	yes	no

Corrosion remains one of the prime problems with Split Set stabilisers since protection of the outer surface of the dowel is not feasible. Galvanising the tube helps to reduce corrosion, but is probably not a preventative measure which can be relied upon for long term applications in aggressive environments. Stainless steel Split Set stabilisers are now available in some sizes.

12.3.3 'Swellex' dowels

Developed and marketed by Atlas Copco, the 'Swellex' system is illustrated in Figure 12.9. The dowel, which may be up to 12 m long, consists of a 42 mm diameter tube which is folded during manufacture to create a 25 to 28 mm diameter unit which can be inserted into a 32 to 39 mm diameter hole. No pushing force is required during insertion and the dowel is activated by injection of high pressure water (approximately 30 MPa or 4,300 psi) which inflates the folded tube into intimate contact with the walls of the borehole.

During 1993 the original Swellex dowel was replaced by the EXL Swellex which is manufactured from a high strength but ductile steel. This steel allows significant displacement without loss of capacity. Stillborg (1994), carried out a series of tests in which bolts and dowels were installed across a simulated 'joint' and subjected to tensile loading. In the EXL Swellex dowel tests, opening of the joint concentrates loading onto the portion of the dowel crossing the joint,

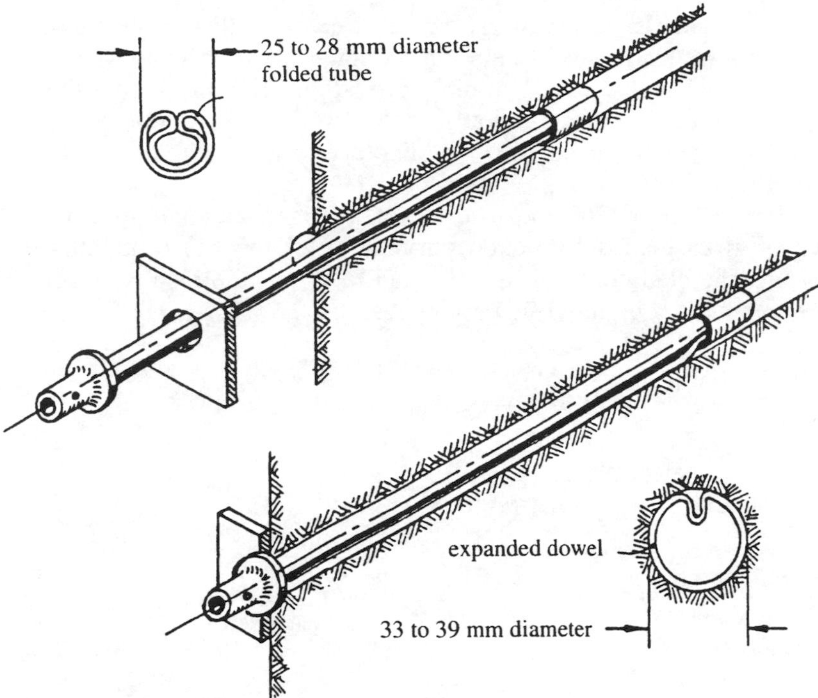

Figure 12.9: Atlas Copco Swellex dowel.

causing a reduction in diameter and a progressive 'de-bonding' of the dowel away from the joint. The ductile characteristics of the steel allows the de-bonded section to deform under constant load until, eventually, failure occurs when the total displacement reaches about 140 mm at a constant load of approximately 11 tonnes. These tests are described in greater detail later in this Chapter.

Corrosion of Swellex dowels is a matter of concern since the outer surface of the tube is in direct contact with the rock. Atlas Copco have worked with coating manufacturers to overcome this problem and claim to have developed effective corrosion resistant coatings.

Speed of installation is the principal advantage of the Swellex system as compared with conventional rockbolts and cement grouted dowels. In fact, the total installation cost of Swellex dowels or Spilt Set stabilisers tends to be less than that of alternative reinforcement systems, when installation time is taken into account. Both systems are ideal for use with automated rockbolters.

12.4 Load-deformation characteristics

Stillborg (1994) carried out a number of tests on rockbolts and dowels installed across a simulated 'joint', using two blocks of high strength reinforced concrete. This type of test gives a more accurate representation of conditions encountered underground than does a standard 'pull-out' test.

The rockbolts and dowels tested were installed in percussion drilled holes using the installation techniques which would be used in a normal underground mining operation. The installed support systems were then tested by pulling the two blocks of concrete apart at a fixed rate and measuring the displacement across the simulated 'joint'.

The results of Stillborg's tests are summarised in Figure 12.10 which gives load deformation curves for all the bolts and dowels tested. The configuration used in each test and the results obtained are summarised on the following pages.

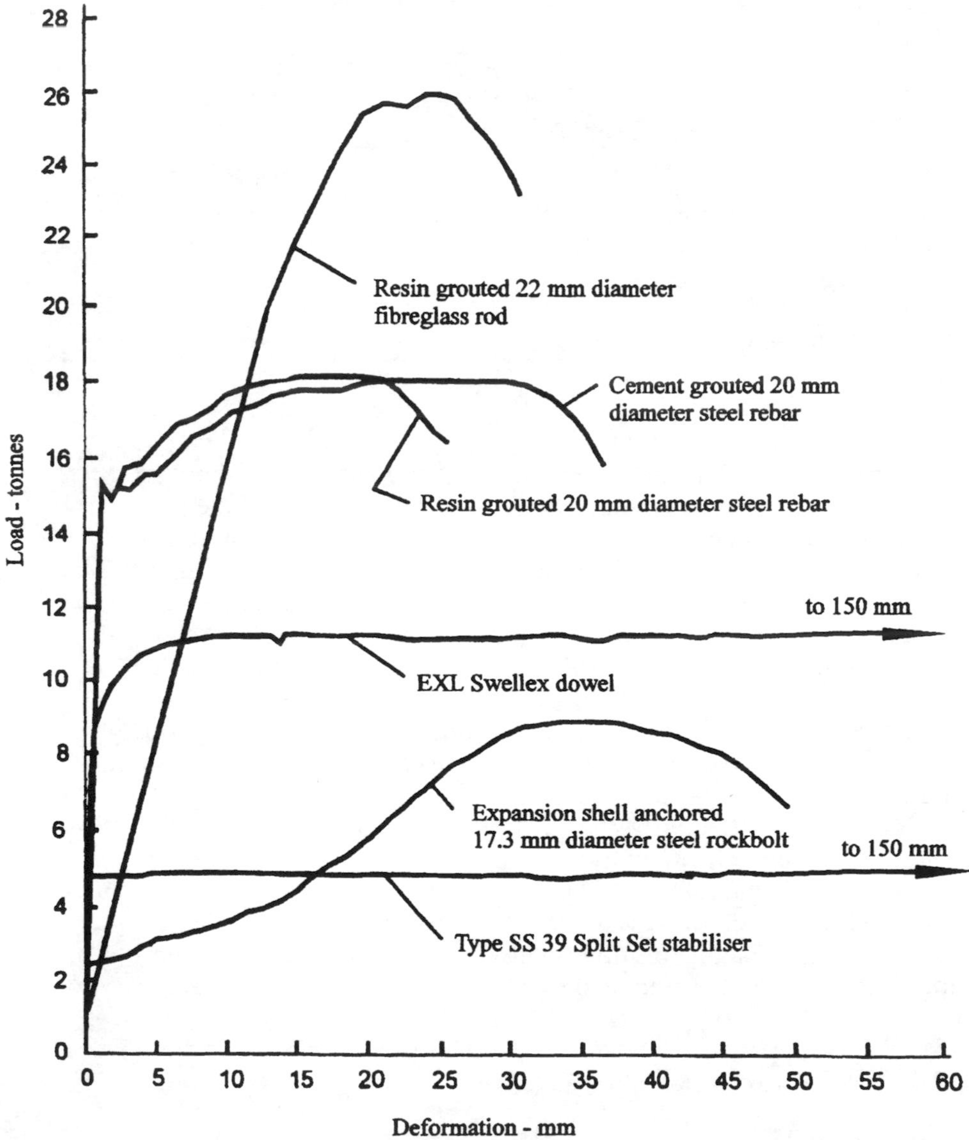

Figure 12.10: Load-deformation results obtained by Stillborg in tests carried out at Luleå University in Sweden. High strength reinforced concrete with a uniaxial compressive strength of 60 MPa was used for the test blocks and holes were drilled with a percussion rig to simulate in situ rock conditions.

1. *Expansion shell anchored rockbolt*

   Steel rod diameter: 17.28 mm
   Ultimate tensile strength of bolt shank: approximately 12.7 tonnes
   Expansion shell anchor: Bail type three wedge anchor
   Face plate: Triangular bell plate, nut with hemispherical seating
   Bolt pre-load: 2.25 tonnes
   Borehole diameter: 34 mm

   At the pre-load of 2.25 tonnes, no deformation of the face plate.

   At a load of 4 tonnes, the face plate has deformed 9.5 mm and is completely flat, the bolt shank has deformed an additional 3.5 mm giving a total deformation of 13 mm at 4 tonnes load.

   Failure initiates at a load of 8 tonnes and a deformation of 25 mm with progressive failure of the expansion shell anchor in which the cone is pulled through the wedge.

   Maximum load is 9 tonnes at a deformation of 35 mm.

2. *Cement grouted steel rebar*

   Steel bar diameter: 20 mm
   Ultimate tensile strength of steel rebar: 18 tonnes
   Faceplate: flat plate
   Borehole diameter: 32 mm
   Cement grout: 0.35 water/cement ratio grout cured for 11 days

   At a load of 15 tonnes and an elastic deformation of about 1.5 mm, a sudden load drop is characteristic of hot rolled rebar steel.

   Maximum load is 18 tonnes at a deformation of 30 mm.

3. *Resin grouted steel rebar*

   Steel rebar diameter: 20 mm
   Ultimate tensile strength of steel rebar: 18 tonnes
   Faceplate: flat plate
   Borehole diameter: 32 mm

   Resin grout: Five 580 mm long, 27 mm diameter polyester resin cartridges. Curing time 60 minutes. Mixed by rotating rebar through cartridges in the borehole

   At a load of 15 tonnes and an elastic deformation of about 1.5 mm, a sudden load drop is characteristic of hot rolled rebar steel.

   Maximum load is 18 tonnes at a deformation of 20 mm.

   The resin is stronger than the cement grout and local fracturing and bond failure in and near the joint is limited as compared with the cement grouted rebar, leading to a reduced ultimate displacement at rebar failure.

4. *Resin grouted fibreglass rod*

   Fibreglass rod diameter: 22 mm
   Ultimate tensile strength of fibreglass rod: 35 tonnes
   Faceplate: special design by H. Weidmann AG. Switzerland (see margin drawing - after Stillborg)
   Borehole diameter: 32 mm

Special faceplate and nut for fibreglass rod designed and manufactured by H. Weidmann AG, Switzerland

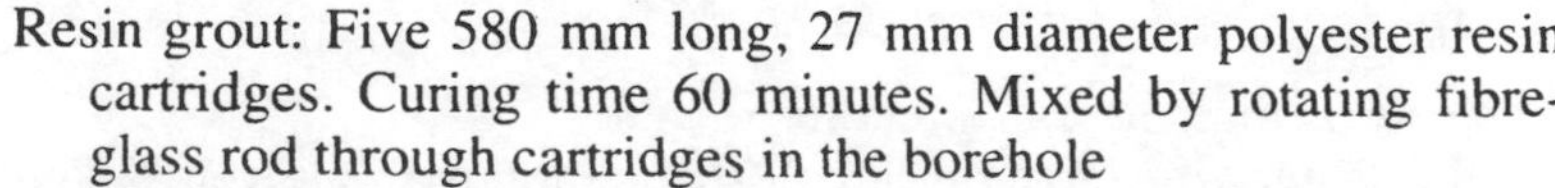

Resin grout: Five 580 mm long, 27 mm diameter polyester resin cartridges. Curing time 60 minutes. Mixed by rotating fibreglass rod through cartridges in the borehole

At approximately 1.5 tonnes load, failure of the fibreglass/resin interface initiates and starts progressing along the rod. As bond failure progresses, the fiberglass rod deforms over a progressively longer 'free' length.

General bond failure occurs at a load of approximately 26 tonnes and a deformation of 25 mm.

The ultimate capacity of this assembly is determined by the bond strength between the resin and the fibreglass rod and by the relatively low frictional resistance of the fibreglass.

5. *Split Set stabiliser, type SS 39*

Tube diameter: 39 mm
Ultimate tensile strength of steel tube: 11 tonnes
Faceplate: special design by manufacturer (see Figure 12.8)
Borehole diameter: 37 mm

Dowel starts to slide at approximately 5 tonnes and maintains this load for the duration of the test which, in this case, was to a total displacement of 150 mm.

6. *EXL Swellex dowel*

Tube diameter: 26 mm before expansion
Ultimate tensile strength of steel tube: 11.5 tonnes (before expansion)
Type of face plate: Domed plate (see margin drawing - after Stillborg)
Borehole diameter: 37 mm
Pump pressure for expansion of dowel: 30 MPa

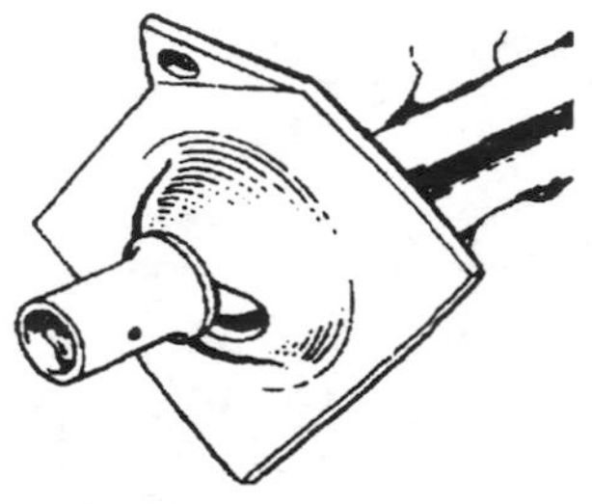

Domed faceplate used by Stillborg in test on EXL Swellex dowel.

At 5 tonnes load the dowel starts to deform locally at the joint and, at the same time, 'bond' failure occurs at the joint and progresses outward from the joint as the load is increased. General 'bond' failure occurs at 11.5 tonnes at a deformation of approximately 10 mm. The dowel starts to slide at this load and maintains the load for the duration of the test which, in this case, was to 150 mm.

13 Cablebolt reinforcement

13.1 Introduction

The move towards larger underground excavations in both mining and civil engineering has resulted in the gradual development of cablebolt reinforcement technology to take on the support duties which exceed the capacity of traditional rockbolts and dowels. A brief review of some typical cablebolt reinforcement applications in underground mining was given in Chapter 11. This Chapter deals with many of the hardware issues which are critical in the successful application of cablebolts in underground hard rock mining, and with factors that affect the bond strength and capacity of cablebolts.

13.2 Cablebolt hardware

The earliest known use of cablebolt reinforcement in underground mining was at the Willroy mine in Canada (Marshall, 1963) and at the Free State Geduld Mines Ltd. in South Africa (Thorn and Muller, 1964). Extensive development of cablebolt reinforcement technology occurred during the 1970s with major contributions being made by mining groups in Australia, Canada and South Africa. The use of cablebolt reinforcement in cut and fill and large non-entry stopes, described in Section 11.7 of Chapter 11, played a crucial role in the development of cablebolt technology. Figure 13.1, after Windsor (1992), gives a summary of some of the cablebolt hardware which has been developed to meet various mining requirements.

Early cablebolts were generally made from discarded winder rope but this practice was discontinued, because of the time-consuming de-greasing process required to make these ropes suitable for grouting into boreholes. Straight, 7 mm diameter, pre-stressing wires were used in Australia in the mid-1970s and are described in papers by Clifford (1974), Davis (1977), Fuller (1981) and Jirovec (1978). The first use of seven wire strands, where the individual wires are spun helically around a central straight 'kingwire' into a single cablebolt, is thought to have been at Broken Hill in Australia (Hunt and Askew, 1977).

Reviews by Fabjanczyk (1982) and Fuller (1984) showed that, where plain strand cablebolts were used in underground mining, almost all failures were associated with the rock stripping off the cablebolts. Very few cases of broken cablebolts were reported, suggesting that the weakest component in the cablebolt reinforcement system is the bond between the grout and the cablebolt. This has been confirmed by extensive laboratory and field tests carried out by Queen's and Laurentian Universities in Canada (Kaiser et al., 1992).

In an attempt to remedy the problem failure of the bond between the steel wires and the grout, various types of barrel and wedge or swaged anchors were used, as shown in Figure 13.1 These were

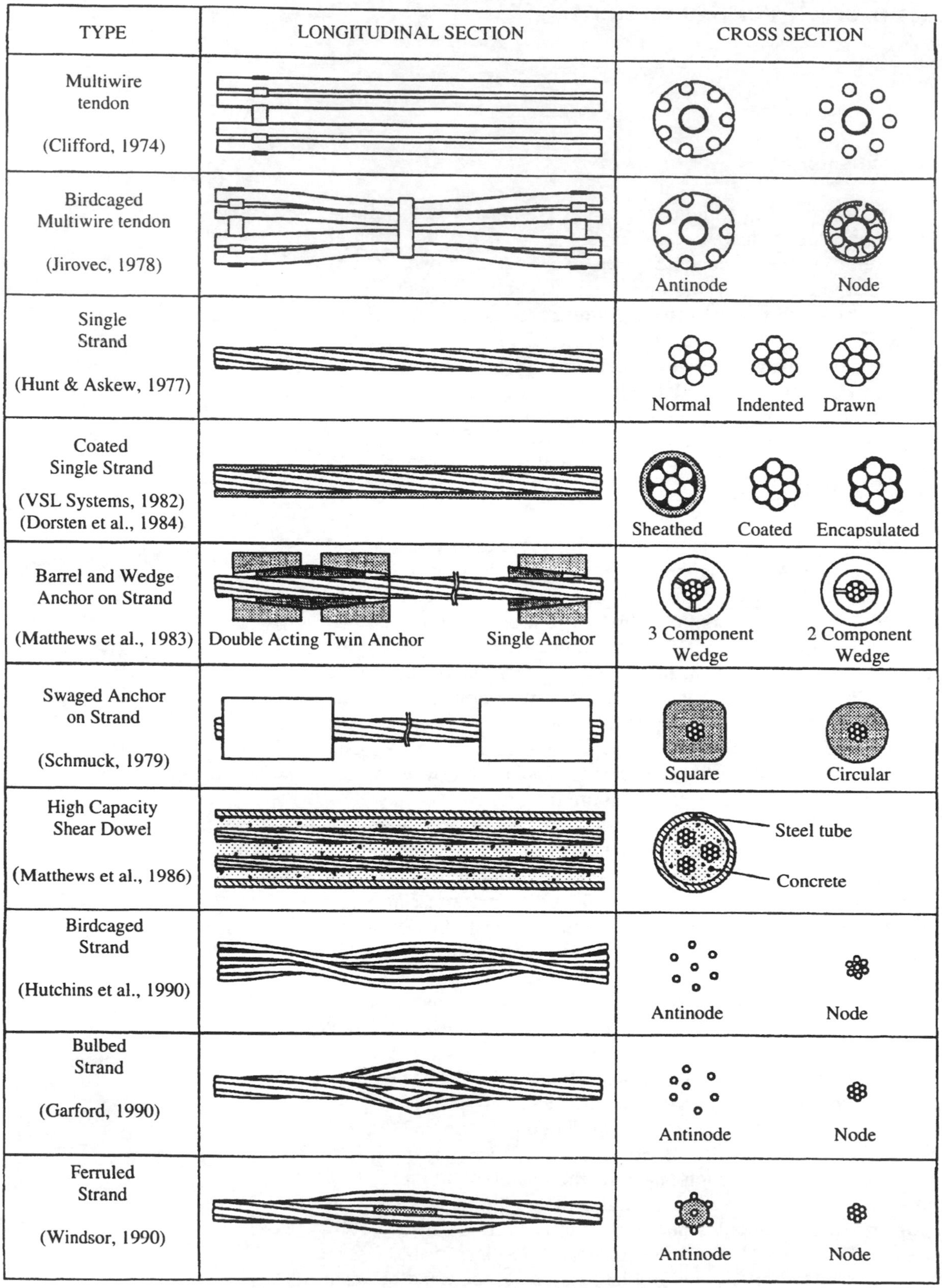

TYPE	LONGITUDINAL SECTION	CROSS SECTION
Multiwire tendon (Clifford, 1974)		
Birdcaged Multiwire tendon (Jirovec, 1978)		Antinode; Node
Single Strand (Hunt & Askew, 1977)		Normal; Indented; Drawn
Coated Single Strand (VSL Systems, 1982) (Dorsten et al., 1984)		Sheathed; Coated; Encapsulated
Barrel and Wedge Anchor on Strand (Matthews et al., 1983)	Double Acting Twin Anchor; Single Anchor	3 Component Wedge; 2 Component Wedge
Swaged Anchor on Strand (Schmuck, 1979)		Square; Circular
High Capacity Shear Dowel (Matthews et al., 1986)		Steel tube; Concrete
Birdcaged Strand (Hutchins et al., 1990)		Antinode; Node
Bulbed Strand (Garford, 1990)		Antinode; Node
Ferruled Strand (Windsor, 1990)		Antinode; Node

Figure 13.1: Summary of the development of cablebolt configurations. After Windsor (1992).

superseded by the development of the simpler and cheaper 'birdcage' cablebolt at Mount Isa Mines in Australia in 1983 (Hutchins et al., 1990).

In most underground hard rock mining applications today, plain seven strand cablebolt or modified cablebolts (birdcage, ferruled, nutcase or bulbed strand) are used for typical cablebolt reinforcement systems. These cablebolts are generally cement grouted into boreholes, either singly or in pairs, and are generally untensioned since they are either installed before stoping commences or sequentially during the stoping operation.

In large civil engineering applications such as underground powerhouse caverns, the cablebolts tend to be grouted into a corrugated plastic sleeve for corrosion protection and the whole assembly is then grouted into the hole. In most cases a 2 to 3 m long grout anchor is formed, at the end of the hole, and allowed to set. The cablebolt is then tensioned and the remainder of the borehole is filled with grout.

13.3 Cablebolt bond strength

The forces and displacements associated with a stressed cablebolt grouted into a borehole in rock are illustrated in Figure 13.2.

As the cablebolt pulls out of the grout, the resultant interference of the spiral steel wires with their associated grout imprints or flutes causes radial displacement or dilation of the interface between the grout and the cablebolt. The radial dilation induces a confining pressure which is proportional to the combined stiffness of the grout and

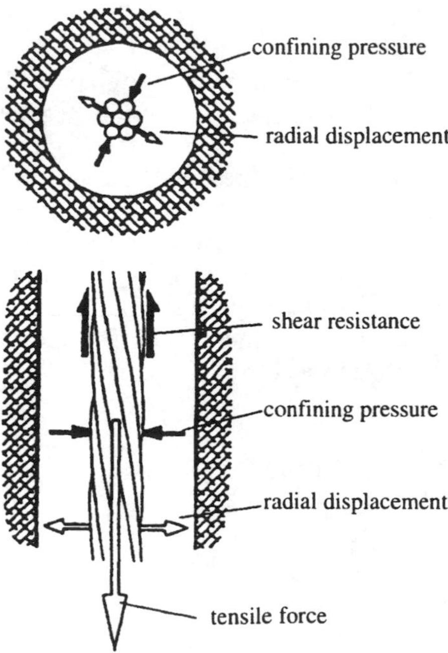

Figure 13.2: Forces and displacements associated with a stressed cablebolt grouted into a borehole in rock.

the rock surrounding the borehole. The shear stress, which resists sliding of the cablebolt, is a product of the confining pressure and the coefficient of friction between the steel wires and the grout. Shear strength, therefore, increases with higher grout strength, increases in the grout and the rock stiffness and increases in the confining stresses in the rock after installation of the cablebolt. Conversely, decrease in shear strength can be expected if any of these factors decrease or if the grout crushes.

Theoretical models of the behaviour of this rock/grout/cablebolt system have been developed by Yazici and Kaiser (1992), Kaiser et al. (1992), Hyett et al., (1992). The first of these models has been incorporated into a program called CABLEBND* (Diederichs et al., 1993) which, when run in conjunction with a companion program called CSTRESS*, predicts cablebolt bond strength for design purposes.

A particularly important aspect of these theoretical models is the influence of stress change in the surrounding rock mass. When the cablebolt is grouted into a borehole prior to mining of a stope, the stresses in the rock can change significantly when stoping commences. In some locations, such as the hanging wall of a stope, the stresses in the rock surrounding the borehole may drop to relatively low levels. These stress reductions may significantly reduce the confining stress acting at the cablebolt to grout interface and hence reduce the shear strength of this interface. Evidence of this process can be seen in many mine stopes where cablebolts, from which the rock has been cleanly stripped, show few signs of distress.

Figure 13.3 gives the results of an analysis using the programs CSTRESS and CABLEBND. The contours in this figure indicate reductions in the cablebolt bond strength in the hanging wall of a stope. Note that in that lower part of the hanging wall, reductions of 50% in cablebolt bond strength, as compared with the initial design strength, are predicted.

Further results obtained from the program CABLEBND are given in Figures 13.4 and 13.5. These give typical cablebolt bond strengths for different values of rock stiffness, stress change magnitudes and grout water/cement ratios. These calculations assume a single 15.2 mm plain cablebolt grouted into a hole of approximately 53 mm diameter. Figure 13.4 shows that the stress change in the rock mass in which the cablebolt is grouted has a significant effect upon the cablebolt bond strength, particularly for stiff rocks (Er = 70 to 90 GPa).

Figure 13.5 shows the importance of grout quality on the cablebolt bond strength, a topic discussed in more detail below.

13.4 Grouts and grouting

The question of grout quality has always been a matter of concern in reinforcement systems for underground construction. One of the critical factors in this matter has been the evolution of grout pumps capable of pumping grouts with a low enough water/cement ratio (by

---

*Available from Program Requests, Geomechanics Research Centre, F217, Laurentian University, Ramsey Lake Road, Sudbury, Ontario, Canada P3E 2C6, Tel. 1 705 675 1151 ext. 5075, Fax 1 705 675 4838.

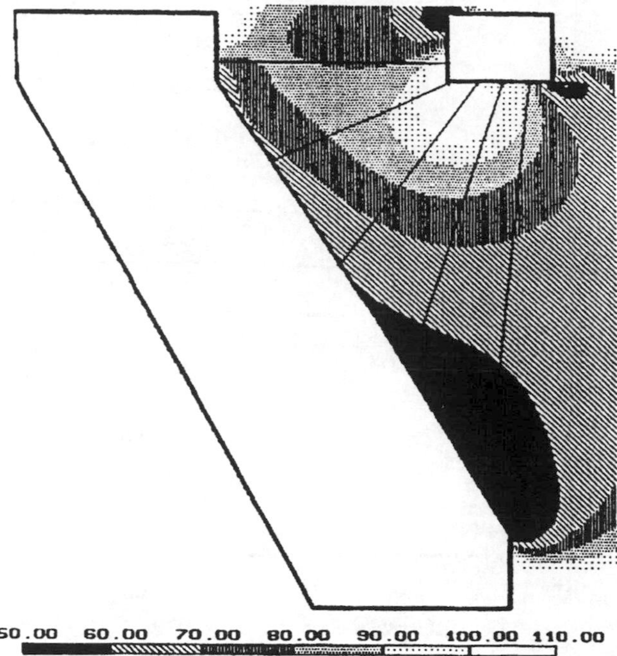

Figure 13.3: Predicted reductions in cablebolt bond strength in the hanging wall of a stope. The models CSTRESS and CABLEBND were used for this analysis (Diederichs et al., 1993).

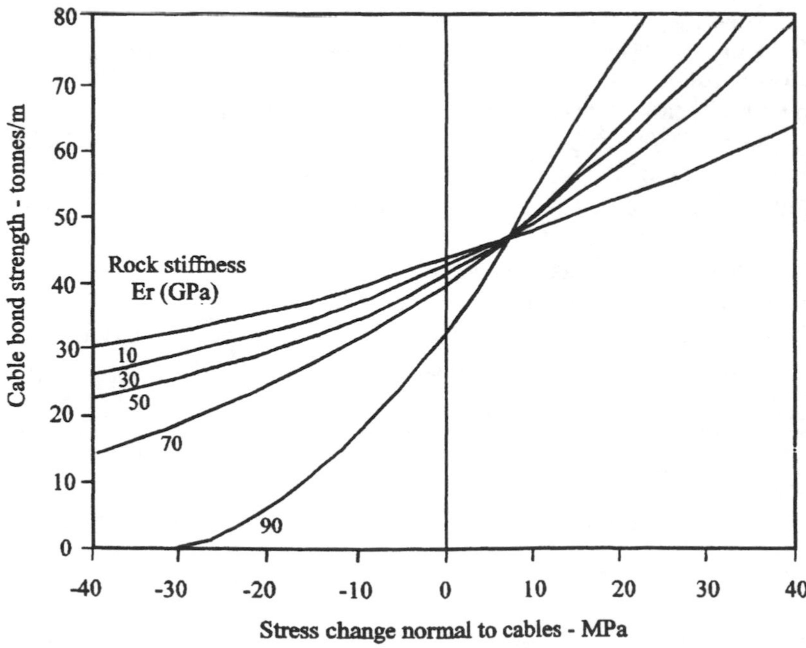

Figure 13.4: Typical cablebolt bond strength values for a range of rock stiffness values and changes in normal stress.

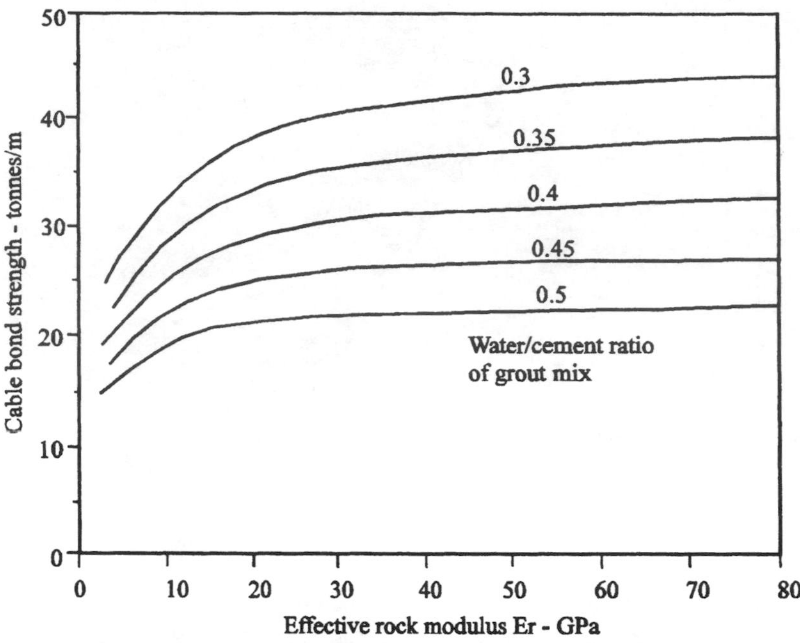

Figure 13.5: Typical cablebolt bond strengths for a range of rock mass stiffness values and different grout water/cement ratios.

weight) to achieve adequate strengths. Fortunately, this problem has now been overcome and there is a range of grout pumps on the market which will pump very viscous grouts and will operate reliably under typical underground mine conditions.

The results of a comprehensive testing programme on Portland cement grouts have been summarised by Hyett et al. (1992) and Figures 13.6, 13.7 and 13.8 are based upon this summary. Figure 13.7 shows the decrease in both 28 day uniaxial compressive strength and deformation modulus with increasing water/cement ratio. Figure 13.8 gives Mohr failure envelopes for three water/cement ratios.

These results show that the properties of grouts with water/cement ratios of 0.35 to 0.4 are significantly better than those with ratios in excess of 0.5. However, Hyett et al. found that the scatter in test results increased markedly for water/cement ratios less than 0.35. The implication is that the ideal water/cement ratio for use with cablebolt reinforcement lies in the range of 0.35 to 0.4.

The characteristics of grouts with different water/cement ratios are described as follows (after Hyett et al., 1992):

w/c ratio	Characteristics at end of grout hose	Characteristics when handled
< 0.30	Dry, stiff sausage structure.	Sausage fractures when bent. Grout too dry to stick to hand. Can be rolled into balls.
0.30	Moist sausage structure. 'Melts' slightly with time.	Sausage is fully flexible. Grout will stick to hand. Easily rolled into wet, soft balls.
0.35	Wet sausage structure. Structure 'melts' away with time.	Grout sticks readily to hand. Hangs from hand when upturned.
0.4	Sausage structure lost immediately. Flows viscously under its own weight to form pancake.	Grout readily sticks to hand but can be shaken free.
0.5	Grout flows readily and splashes on impact with ground.	Grout will drip from hand – no shaking required.

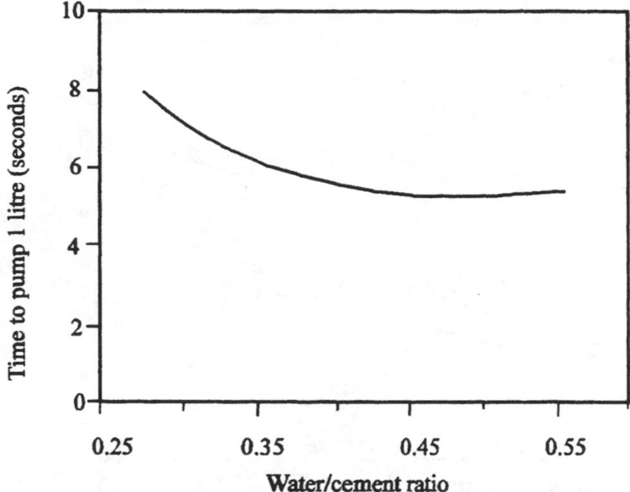

Figure 13.6: Time required to pump one litre of grout with a pump using a helical auger for both mixing and pumping (after Hyett et al., 1992).

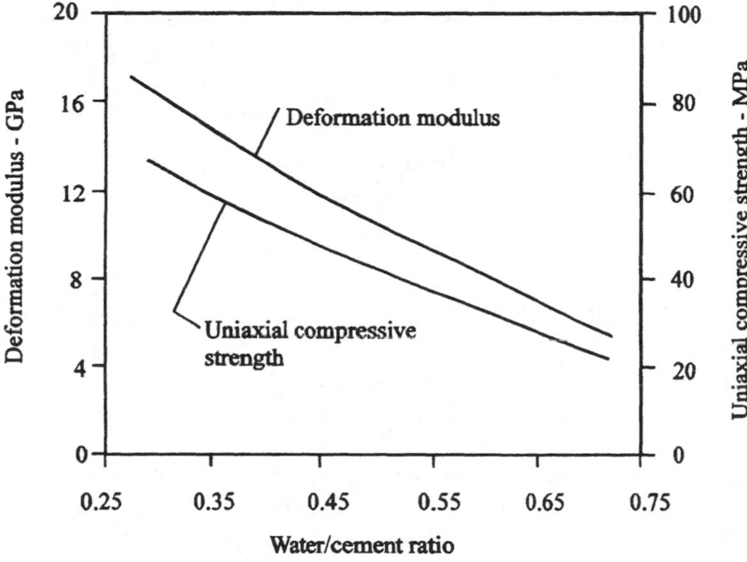

Figure 13.7: Relationship between water/cement ratio and the average uniaxial compressive strength and deformation modulus for grouts tested at 28 days.

13.5 Cablebolt installation

The left hand drawing in Figure 13.9 shows the traditional method of grouting a cablebolt in an uphole. This method will be called the 'breather tube method'. The grout, usually having a water/cement ratio ≥ 0.4, is injected into the bottom of the hole through a large diameter tube, typically 19 mm diameter. The air is bled through a smaller diameter tube which extends to the end of the hole and which is taped onto the cablebolt. Both tubes and the cablebolt are sealed

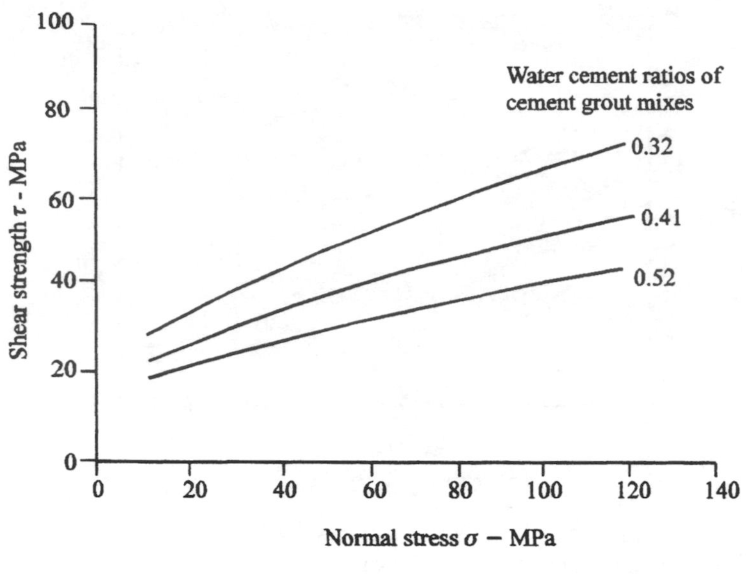

w/c ratio	σ_c MPa	constant m	constant s	Friction angle ϕ°	Cohesion c MPa
0.32	78	3.05	1	24	25
0.41	54	2.14	1	20	19
0.52	38	1.67	1	17	14

Figure 13.8: Mohr failure envelopes for the peak strength of grouts with different water/cement ratios, tested at 28 days.

into the bottom of the hole by means of a plug of cotton waste or of quick setting mortar. As shown, the direction of grout travel is upwards in the hole and this tends to favour a grout column which is devoid of air gaps since any slump in the grout tends to fill these gaps.

Apart from the difficulty of sealing the collar of the hole, the main problem with this system is that it is difficult to detect when the hole is full of grout. Typically, the hole is judged to be full when air ceases to flow from the bleed tube. This may occur prematurely if air is vented into an open joint along the hole. In addition, a void the size of the bleed tube is likely to be left in the grout column. Therefore, it is preferable to stop grouting the borehole only when grout returns along the bleed tube. However, a viscous grout will not flow down a 9 mm bleed tube and so a larger tube is required.

An alternative method, called the 'grout tube method' is illustrated in the right hand drawing in Figure 13.9. In Canada this method, known locally as the 'Malkoski method', has been adopted by some mining groups for use with plain strand single and double cablebolts installed in upholes. In this case a large diameter grout injection tube extends to the end of the hole and is taped onto the cablebolt. The cablebolt and tube are held in place in the hole by a wooden wedge inserted into the hole collar. Note that care has to be taken to avoid compressing the grout tube between the wedge and the cablebolt. Grout is injected to the top of the hole and is pumped

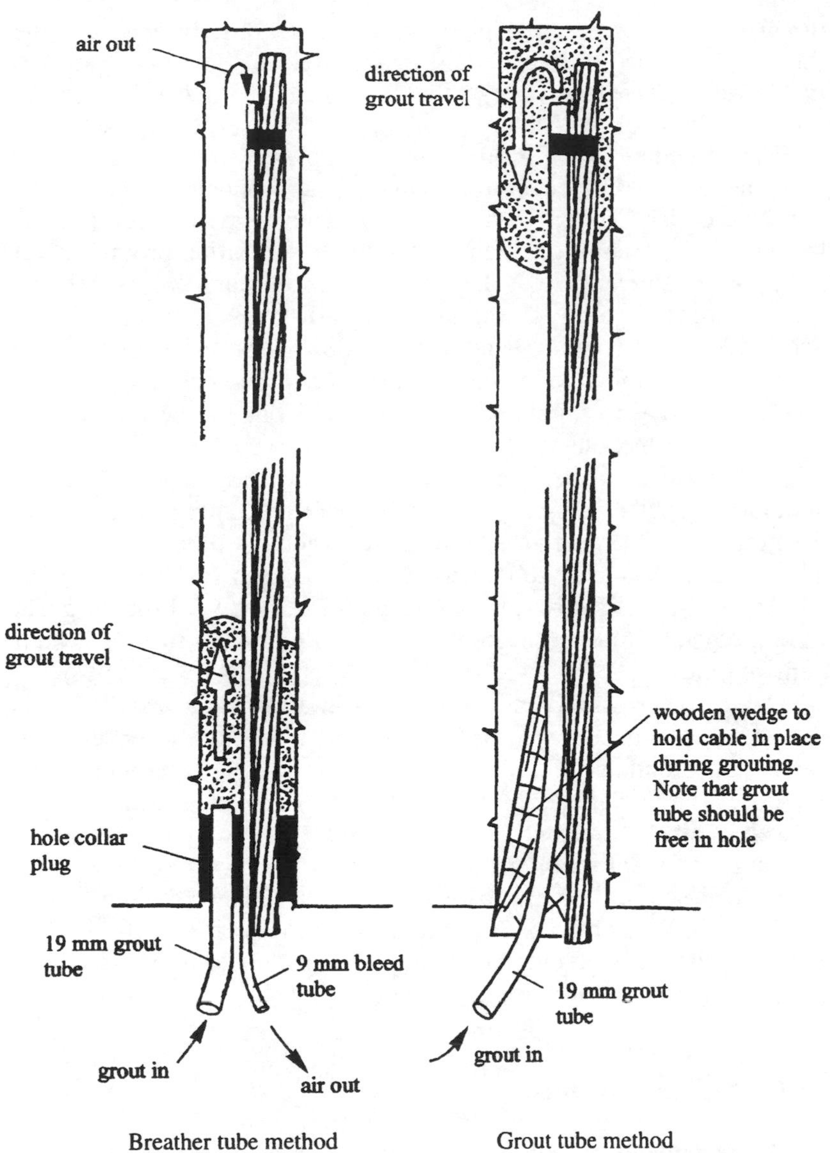

Figure 13.9: Alternative methods for grouting cablebolts into upholes.

down the hole until it appears at the hole collar. If a watery grout appears first at the collar of the hole, grout pumping is continued until a consistently thick grout is observed.

Provided that a very viscous mix is used (0.3 to 0.35 water/cement ratio), the grout will have to be pumped into the hole and there is little danger of slump voids being formed. However, a higher water/cement ratio mix will almost certainly result in air voids in the grout column as a result of slumping of the grout. The principal advantage of this method is that it is fairly obvious when the hole is full of grout and this, together with the smaller number of components required, makes the method attractive when compared with the traditional method for grouting plain strand cablebolts. In addition, the thicker grout used in this method is not likely to flow into fractures in the rock, preferring instead the path of least flow resistance towards the borehole collar.

The procedure used for grouting downholes is similar to the grout tube method, described above, without the wooden wedge in the borehole collar. The grout tube may be taped to the cablebolt or retracted slowly from the bottom of the hole as grouting progresses. It is important to ensure that the withdrawal rate does not exceed the rate of filling the hole so the air voids are not introduced. This is achieved by applying, by hand, a slight downward force to resist the upward force applied to the tube by the rising grout column. Grout of any consistency is suitable for this method but the best range for plain strand cablebolts is between 0.3 and 0.4 water/cement ratio.

Modified cablebolts, such as birdcage, ferruled or bulbed strand, should be grouted using a 0.4 water/cement ratio mix to ensure that the grout is fluid enough to fill the cage structure of these cablebolts. Therefore, the breather tube method must be used for these types of cablebolts, since the grout flow characteristics required by the grout tube method is limited to grouts in the range of 0.3 to 0.35 water/cement ratio.

One of the most critical components in a cablebolt installation is the grout column. Every possible care must be taken to ensure that the column contains as few air voids as possible. In the breather tube method, a large diameter breather tube will allow the return of grout as well as air. When using the grout tube method in upholes, a 0.3 to 0.35 water/cement ration grout will ensure that pumping is required to cause the grout column to flow, and this will avoid slumping of the grout in the borehole. A grout with a water/cement ratio of less than 0.3 should be avoided, since it will tend to form encapsulated air voids as it flows around the cablebolt.

13.6 Modified cablebolts

Modified cablebolts, such as the birdcaged, bulbed or ferruled strand cablebolts illustrated in Figure 13.1, are useful where a reduction in confining stress is likely to cause a reduction in the bond strength of plain strand cablebolts. A typical situation, where this can occur, is illustrated in Figure 13.3 and the significant bond strength reductions associated with reduced confining stresses are shown in Figure 13.4.

In the case of modified cablebolts, the penetration of the grout into the cage structures results in a mechanical interference, which is much less sensitive to confining stress change than the plain strand cablebolt shown in Figure 13.2. Consequently, modified cablebolts can be expected to maintain a high bond strength in situations, such as the stope hanging wall illustrated in Figure 13.3, where significant stress reductions can occur.

Field tests at the Hemlo Golden Giant Mine in Canada gave the following average peak loads for cables with an embedment length of 300 mm (Hyett et al., 1993):

Location	Plain 7 wire cablebolt	Birdcage cablebolt	Nutcase cablebolt*
Hanging wall	15.4 tonnes	27.7 tonnes	30.4 tonnes
Ore	17.9 tonnes	24.2 tonnes	27.6 tonnes

Although these test results are preliminary, they do indicate that there is a substantial increase in bond strength for modified cablebolts as compared with the plain strand cablebolts.

This strength increase can be particularly important in cases where it is not possible to attach a faceplate. Orepasses, drawpoints or non-entry stopes, in which cablebolts are installed from a remote access, are examples of such cases. An example of a suggested application of modified cablebolts in a non-entry stope is illustrated in Figure 13.10. In this example the cablebolt sections close to the hangingwall are modified to compensate for the lack of faceplates.

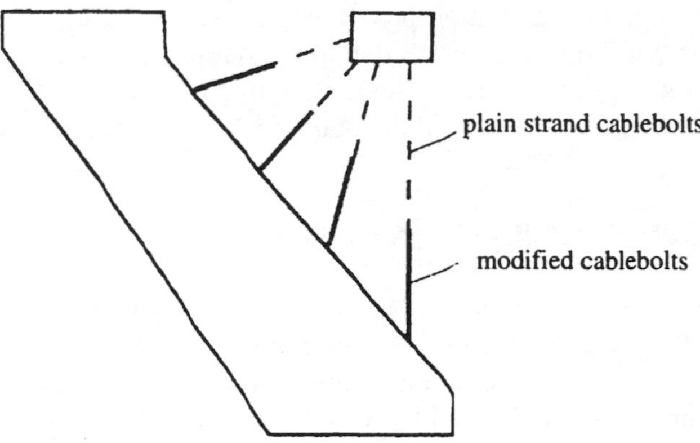

Figure 13.10: Suggested application of modified cablebolts, installed from a remote access, to provide hangingwall support in a non-entry stope.

*A nutcase cable is manufactured by unwinding the cable, threading a series of nuts onto the 'kingwire' and then rewinding the cable. This results in a local flaring of the wires in the vicinity of each nut.

14 The Stability Graph method

14.1 Introduction

Potvin (1988), Potvin and Milne (1992) and Nickson (1992), following earlier work by Mathews et al. (1981), developed the Stability Graph Method for cablebolt design. The current version of the method, based on the analysis of more than 350 case histories collected from Canadian underground mines, accounts for the key factors influencing open stope design. Information about the rock mass strength and structure, the stresses around the opening and the size, shape and orientation of the opening is used to determine whether the stope will be stable without support, stable with support, or unstable even if supported. The method also suggests ranges of cablebolt density when the design is in the realm of 'stable with support'.

14.2 The Stability Graph method

The design procedure is based upon the calculation of two factors, N', the modified stability number which represents the ability of the rock mass to stand up under a given stress condition, and S, the shape factor or hydraulic radius which accounts for the stope size and shape.

14.2.1 *The stability number, N'*

The stability number, N', is defined as

$$N' = Q' \times A \times B \times C \qquad (14.1)$$

where Q' is the modified Q Tunnelling Quality Index
A is the rock stress factor
B is the joint orientation adjustment factor
C is the gravity adjustment factor

The modified Tunnelling Quality Index, Q', is calculated from the results of structural mapping of the rock mass in exactly the same way as the standard *NGI* rock mass classification, except that the stress reduction factor *SRF* is set to 1.00. The system has not been applied in conditions with significant groundwater, so the joint water reduction factor J_w is commonly 1.0. This process is identical to that used earlier in this book for estimating the strength of jointed rock masses (see Equation 8.18 on page 97).

The rock stress factor, A, reflects the stresses acting on the free surfaces of open stopes at depth. This factor is determined from the unconfined compressive strength of the intact rock and the stress acting parallel to the exposed face of the stope under consideration. The intact rock strength can be determined from laboratory testing of the rock or from estimates such as those discussed in Chapter 8. The induced compressive stress is found from numerical modelling or estimated from published stress distributions such as those in Hoek and Brown (1980a), using measured or assumed in situ stress values. The rock stress factor, *A*, is then determined from σ_c/σ_1, the ratio of the intact rock strength to the induced compressive stress on the opening boundary:

$$\begin{aligned} &\text{for } \sigma_c/\sigma_1 < 2 : A = 0.1 \\ &\text{for } 2 < \sigma_c/\sigma_1 < 10 : A = 0.1125\,(\sigma_c/\sigma_1) - 0.125 \\ &\text{and for } \sigma_c/\sigma_1 > 10 : A = 1.0 \end{aligned} \tag{14.2}$$

A plot of the rock stress factor *A*, for different values σ_c/σ_1 is given in Figure 14.1.

The joint orientation adjustment factor, *B*, accounts for the influence of joints on the stability of the stope faces. Most cases of structurally controlled failure occur along critical joints which form a shallow angle with the free surface. The shallower the angle between the discontinuity and the surface, the easier it is for the bridge of intact rock, shown in Figure 14.2, to be broken by blasting, stress or by another joint set. When the angle θ approaches zero, a slight strength increase occurs since the jointed rock blocks act as a beam. The influence of the critical joint on the stability of the excavation surface is highest when the strike is parallel to the free surface, and smallest when the planes are at right angles to one another. The factor, *B*, which depends on the difference between the orientation of the critical joint and each face of the stope, can be found from the chart reproduced in Figure 14.3.

The final factor, *C*, is an adjustment for the effects of gravity. Failure can occur from the roof by gravity induced falls or, from the stope walls, by slabbing or sliding.

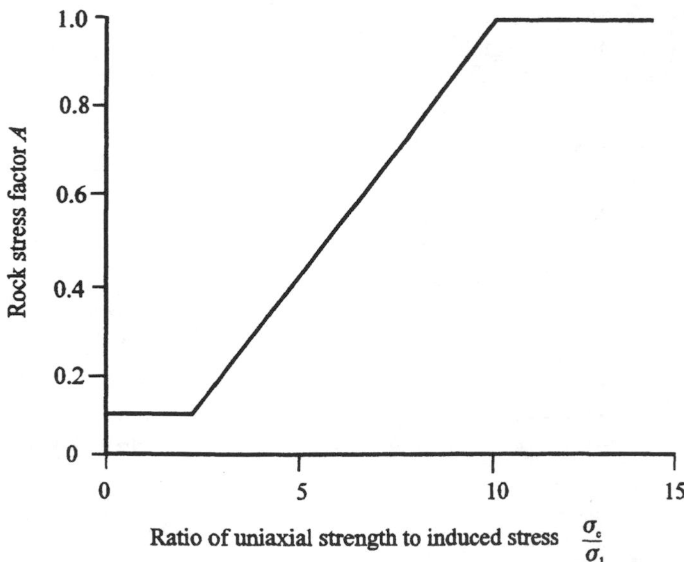

Figure 14.1: Rock stress factor *A* for different values of σ_c/σ_1

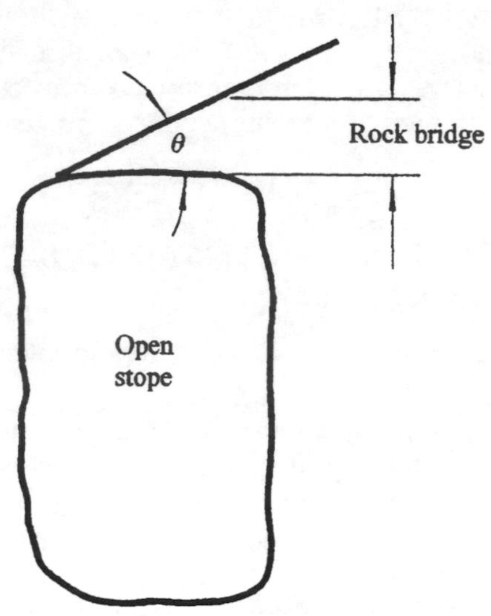

Figure 14.2: Critical joint orientation with respect to the opening surface (After Potvin, 1988).

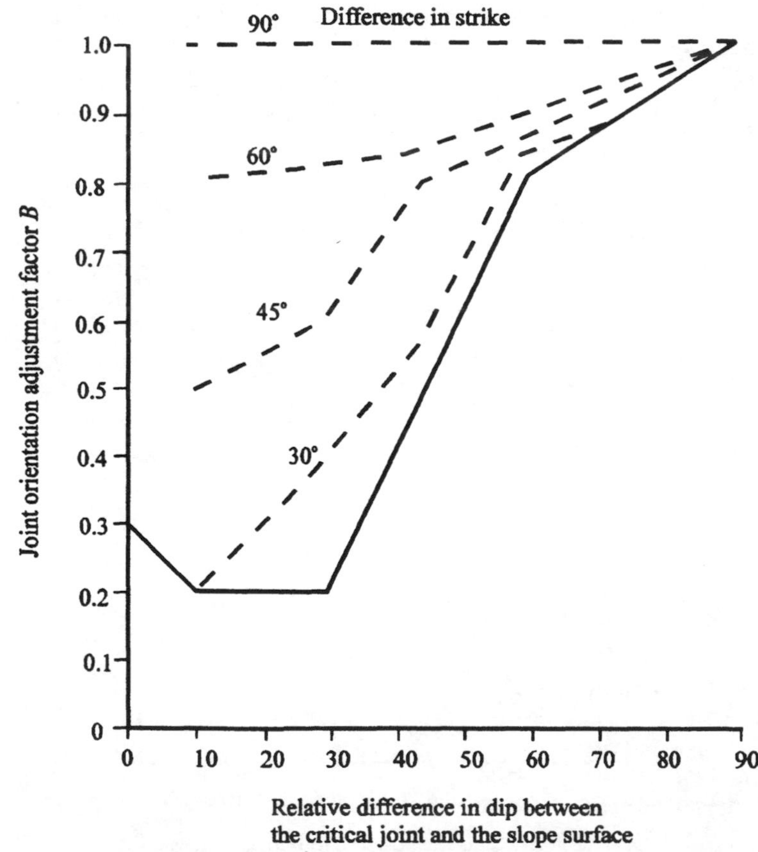

Figure 14.3: Adjustment factor, B, accounting for the orientation of the joint with respect to the stope surface (After Potvin, 1988).

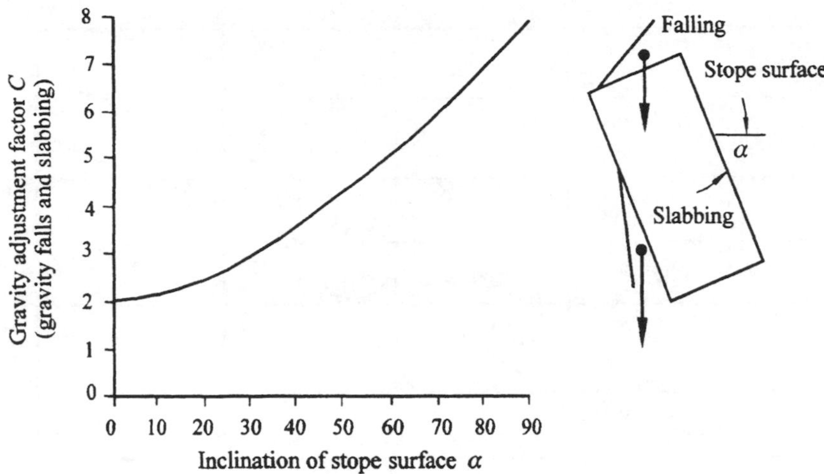

Figure 14.4: Gravity adjustment factor C for gravity falls and slabbing. After Potvin (1988).

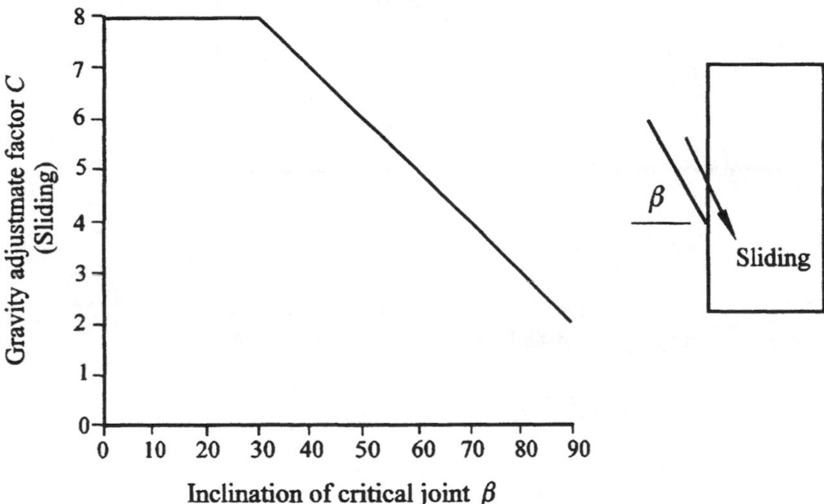

Figure 14.5: Gravity adjustment factor C for sliding failure modes. After Potvin (1988).

Potvin (1988) suggested that both gravity induced failure and slabbing failure depend on the inclination of the stope surface α. The factor C for these cases can be calculated from the relationship, $C = 8 - 6\cos\alpha$, or determined from the chart plotted in Figure 14.4. This factor has a maximum value of 8 for vertical walls and a minimum value of 2 for horizontal stope backs.

Sliding failure will depend on the inclination β of the critical joint, and the adjustment factor C is given in Figure 14.5.

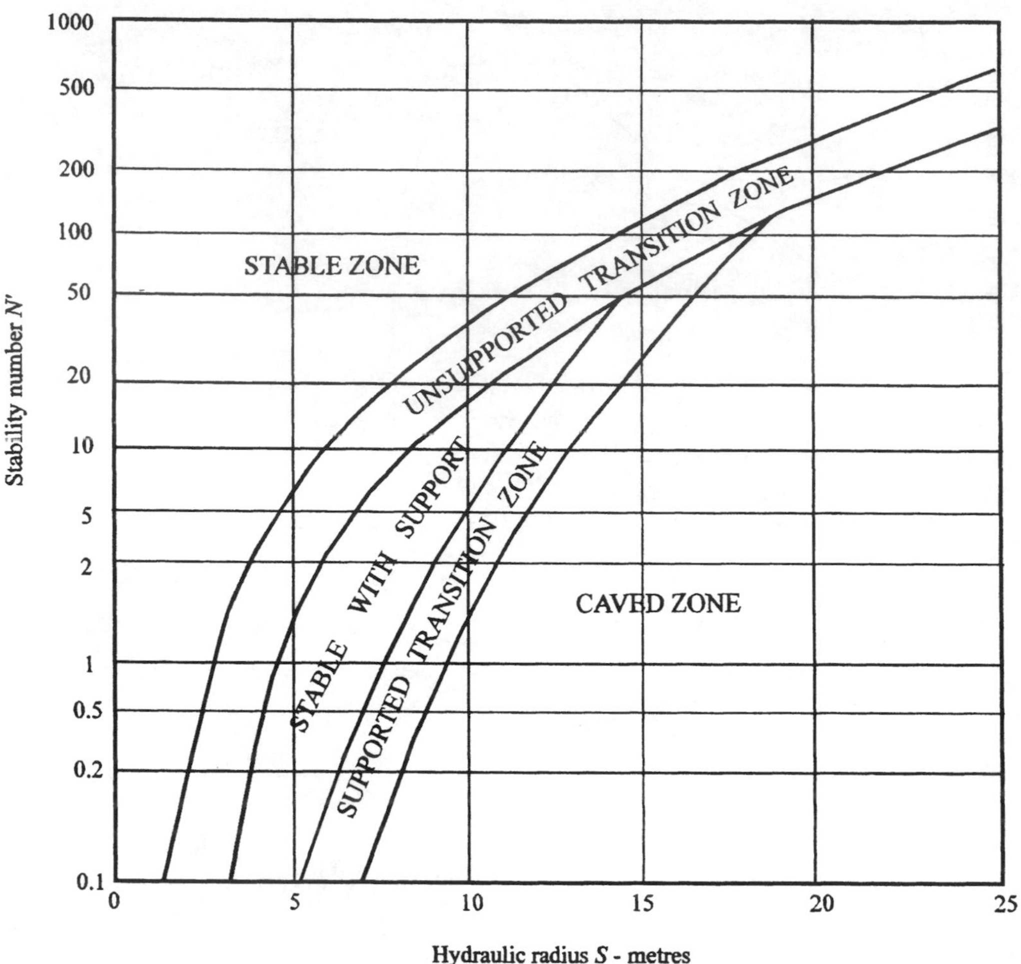

Figure 14.6: Stability graph showing zones of stable ground, caving ground and ground requiring support. After Potvin (1988), modified by Nickson (1992).

14.2.2 *The shape factor, S*

The hydraulic radius, or shape factor, for the stope surface under consideration, is calculated as follows:

$$S = \frac{\text{Cross sectional area of surface analysed}}{\text{Perimeter of surface analysed}} \quad (14.3)$$

14.2.3 *The stability graph*

Using the values of N', the stability number, and S, the hydraulic radius, the stability of the stope can be estimated from Figure 14.6. This figure represents the performance of open stopes observed in many Canadian mines, as tabulated and analysed by Potvin (1988) and updated by Nickson (1992).

14.3 Cablebolt design

Where the stability analysis indicates that the stope requires support, the chart given in Figure 14.7 can be used as a preliminary guide for the cablebolt density. In this chart, the cablebolt density is related to the frequency of jointing through the block size (parameters RQD/J_n) and the hydraulic radius of the opening; both must be considered to get an idea of the relative size of the blocks. Of the three design envelopes shown in this figure, the one used should be based both on the use of the opening and on experience with cablebolt support at the site. At the start of a project, the designer should consider using the more conservative envelopes.

Potvin et al. (1989) noted that there is a great deal of scatter in the data used in deriving Figure 14.7, reflecting the trial and error nature of current cablebolt design. They also stated that cable bolts are not likely to be effective when the relative block size factor, $(RQD/J_n)/$ Hydraulic radius, is less than 0.75 and when the cable bolt density is less than 1 bolt per 10 square metres at the opening boundary.

The length of the cablebolts must be such that they are anchored far enough into undisturbed ground for the anchor to be effective. Potvin et al. suggested that a rough guideline for design is that the length of the cablebolt should be approximately equal to the span of the opening. They found that cablebolts are generally not successful in stabilising very large stopes.

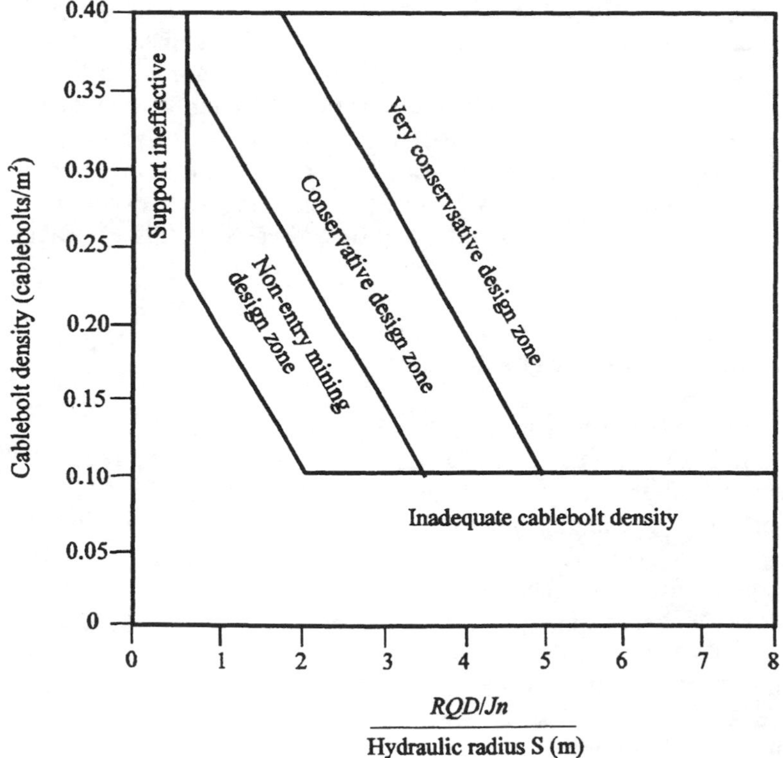

Figure 14.7: Cablebolt density design chart. After Potvin and Milne (1992).

Potvin et al. (1989) suggested that the design of the cablebolts must include consideration of the potential failure mechanism. Where the failure is predicted to be by sliding, the cablebolt should be inclined at 17° to 27° to the plane on which sliding is likely to occur. The most favourable orientation of cable bolts supporting a slabbing failure is perpendicular to the foliation.

14.4 Discussion of the method

Potvin and Milne (1992) warn that the use of the design charts must be limited to the conditions similar to those encountered in the mines used as case histories in the development of the empirical data base. Anomalous geological conditions such as faults, shear zones, dykes or waste inclusions, the creation of a slot or brow within the stope and poor cablebolt installation can all lead to inaccurate results. In addition the cablebolts must cover the excavation surface fully, since the support design is based upon the assumption that the cables form a continuous zone of reinforced rock surrounding the opening.

Practical observations suggest that the main area of uncertainty in using the method lies in the density of jointing in the rock mass. Where the number of joints and other discontinuities per unit volume of rock is highly variable, the value of Q' will be open to question. Under these conditions, the design derived from the stability graph method should be regarded as a first step in the design process and local adjustments to the design will have to be made, depending upon the conditions observed in the stope.

The quality of the cable bolt installation is another variable which has to be recognised when using this method. Where uncertainty on the effectiveness of grouting exists, a conservative approach has to be adopted. In addition, the use of modifying elements such as plates or birdcaged cable bolts has not been included in the design method, perhaps because these items were not used a great deal at the time of the development of the charts. With time and increasing experience, it is likely that these shortcomings will be addressed in this empirical design method.

14.5 Worked stability graph example

A 15 m thick (hangingwall to footwall) orebody is located at a depth of 500 m below surface and is to be mined by open stoping methods. Access is from the hangingwall and the option exists to fan cablebolt support into the hangingwall from cable stub drifts. Details of the structural geology of the rock mass and the Q' classification are given in the following sections.

Stope design, using the stability graph method, is an iterative process. To start with, reasonable stope dimensions depending upon drilling access, practical mining considerations, and economics, should be proposed, as illustrated in the margin sketch. In this example, the full width of the orebody (15 metres) will be mined in a single stope, and the drilling sublevels are planned for every 25 metre interval of depth, with over and undercuts every 100 metres. The stability graph design procedure is then carried out for these dimensions. This analysis indicates the stability of the proposed stope, and if the dimensions have to be altered, further analyses should be carried on the new dimensions. The procedure is iterated until a satisfactorily stable design is achieved.

In this case, the geometry of the orebody suggests that the support should take the form of rings of cable bolts, installed from the access drifts. These rings should extend up from each drift to support the back of each stope, and out from the end of the drift into the hanging wall rock mass, thus creating a stable 'buttress' at each drift. Also, if the design indicates the need for additional support, cables should be installed from hangingwall access drifts to provide adequate coverage.

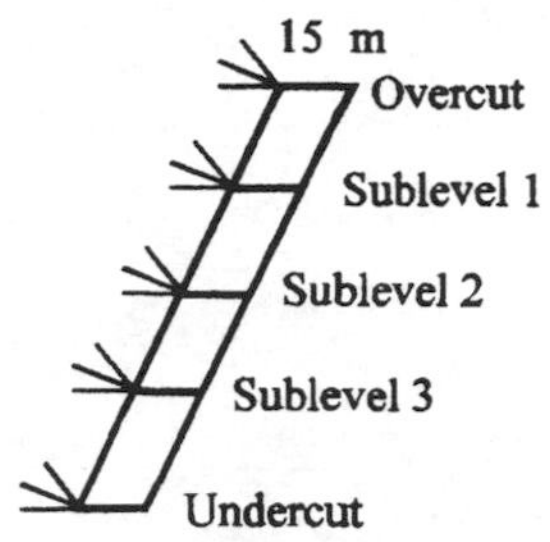

Cablebolts installed from access drifts

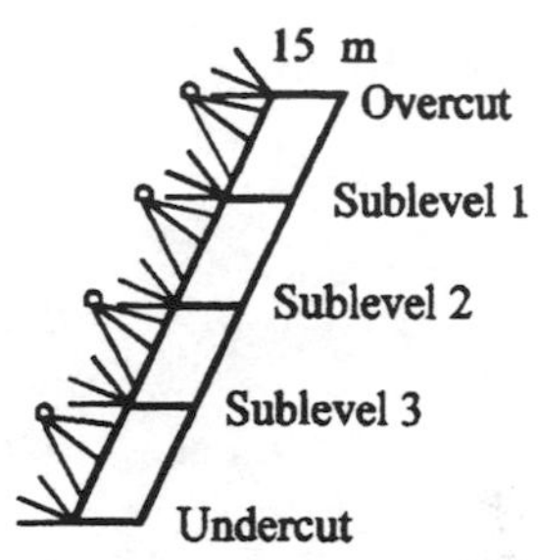

Additional cablebolts installed from hanging-wall access drifts if required

14.5.1 *Structural geology*

The orebody strikes east-west and dips at 65° to the north. Extensive borehole core logging and underground mapping have been carried out and a total of 1250 features have been recorded. Analysis of this structural geology information by means of the program DIPS indicates that the rock mass contains 5 joint sets which are described in Table 14.1.

14.5.2 *Q' classification*

The data collected from the geological mapping is used to calculate the modified Tunnelling Quality index, Q' as defined by Equation 8.18 on page 97. Hence $Q' = RQD / J_n \times J_r / J_a$.

The average *RQD* value for the rock mass was found to be 78, with a range from 70 to 86.

Based upon an inspection of the rock mass in the shaft and development excavations, it was decided that not all five joint sets occurred at all locations and that a reasonable description of the jointing is 'three sets plus one random set'. Table 4.6.2G, gives the value of the joint set number for this description as $J_n = 12$.

Table 14.1: Structural geology for the example mine.

Set	Dip°	Dip direction°	Description
A	64±10	009±20	Planar, smooth to medium roughness. Mica or calcite infilling with some gouge zones and sericite. Spacing 10-30 cm. Joints parallel the orebody hangingwall.
B	84±7	098±24	Slightly rough to rough with no infilling. Spacing 35-45 cm. Joints are perpendicular to the orebody.
C	15±9	180±40	Poorly developed but continuous over several metres in places. Rough with calcite or gouge infilling. Widely spaced.
D	47±9	095±9	Striking parallel to B joint, but with shallower dip. Planar, smooth to medium rough, with no infilling. Spacing 50 cm.
E	45±8	271±13	Not evident very often; classed as random. Smooth to medium rough with little or no infilling.

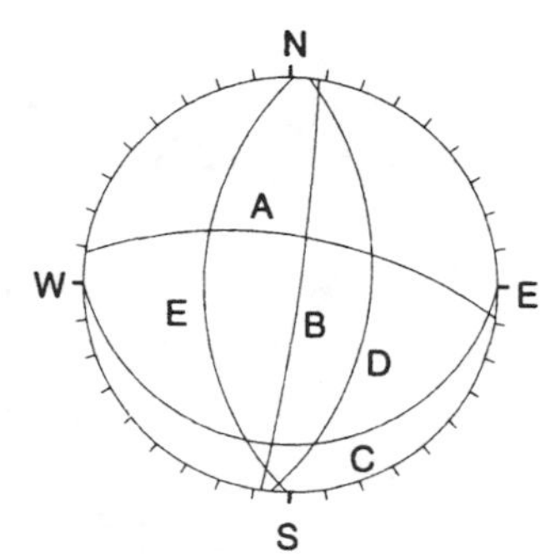

The joint roughness number J_r was found to vary between 1 (smooth planar in Table 4.6.3F) and 2 (smooth undulating in Table 4.6.3C). Similarly, the joint alteration number J_a was found to vary between 1 (unaltered joint walls, surface staining only in Table 4.6. 4B) and 2 (slightly altered joint walls with discontinuous coatings of mica or sericite in Table 4.6.4C). The values chosen for inclusion in the evaluation of Q' were dependent upon the location of the stope being designed and the joint set or sets considered to be most important at that location.

Values of Q', together with laboratory evaluations of the intact rock strength σ_c, Young's modulus E, and Poisson's ratio ν, are:

Location	Q'	σ_c (MPa)	E (GPa)	ν
Hangingwall	2.4	70	40	.25
Ore zone (Stope back)	6.3	100	53	.10
Footwall	5.1	175	55	.21

14.5.3 *Preliminary stope design*

The preliminary stope design will be based upon stope dimensions of a stope back span of 15 m and a 25 m stope height. The assessment of the stability and the third stope dimension (strike width in this case) then depends on the estimates of the factors A, B and C, included in Equation 14.1.

Factor A, the influence of the mining induced stresses, is found from Equation 14.2 from the ratio of the intact rock strength to the induced compressive stress, σ_c/σ_1. The intact rock strength is discussed above, and the induced compressive stress can be estimated from a consideration of the in situ stresses and the proposed stope geometry. The in situ stresses are listed in the following table, and the orientations plotted on the lower hemisphere projection shown in the margin sketch.

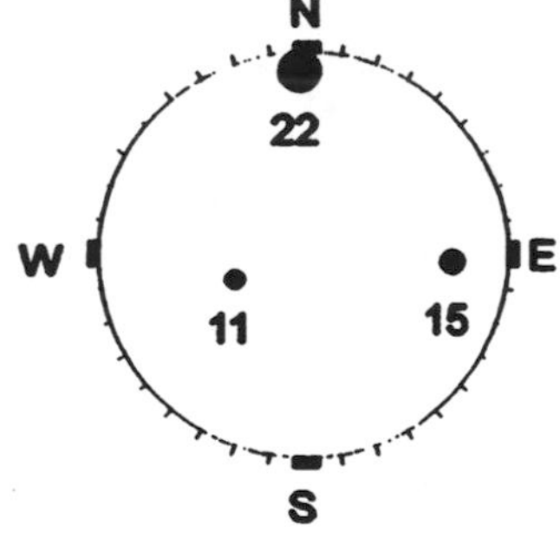

	Trend°	Plunge°	Magnitude (MPa/m depth)	Magnitude at 500m depth (MPa)
σ_1	358	10	0.0437	21.9
σ_2	093	28	0.0299	15
σ_3	250	60	0.0214	10.7

A preliminary estimate of the induced compressive stress on each portion of the stope boundary can be obtained from simple elastic numerical modelling. As discussed above, the stope back and hangingwall dimensions have been established from practical mining considerations. The stability graph can then be used to determine a reasonable value for the stope width.

A PHASES analysis of a 15 m span, 25 m high stope gives the contours of the major principal stress reproduced in Figure 14.8. From this plot, the induced compressive stress on the back of the stope is found to be about 30 MPa, and on the hangingwall it is less than 5 MPa. An unconfined compressive strength of $\sigma_c = 100$ MPa is assumed for the ore in the stope back and $\sigma_c = 70$ MPa is assumed for the hangingwall rock. Therefore, the respective ratios of σ_c/σ_1 are

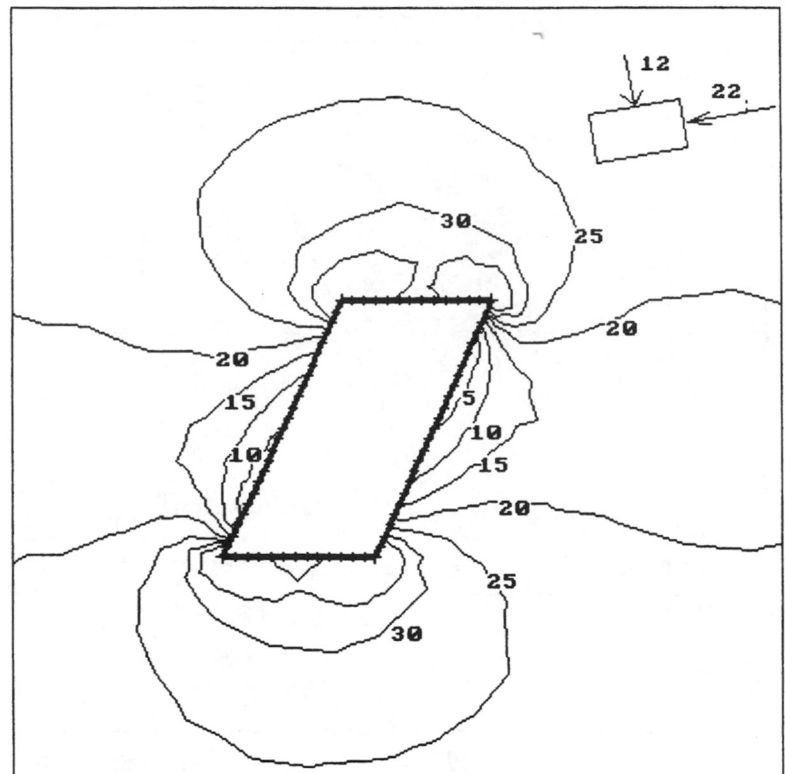

Figure 14.8: Contours of maximum principal stress σ_1 induced in the rock surrounding a 15 m span, 25 m high stope. The in situ stresses acting on the stope are assumed to be 22 MPa (inclined at 10°) and 12 MPa as shown in the figure.

approximately 3.3 and 14. Using these values, the rock stress factor can be calculated from Equation 14.2, giving $A = 0.25$ for the stope back and $A = 1$ for the hangingwall.

The factor B is used to account for the influence of the joint orientation on the stope stability. The most critical joint, influencing the stability of the stope boundary, is generally the one closest to parallel to the boundary. For this example, the critical joint sets for the various components of the stope boundary are listed below, together with the values of B, found from Figure 14.9.

	Joint Set	Difference in strike°	Difference in dip°	Factor B
Stope hangingwall	A	0	0	0.3
Stope ends	B	0	0	0.3
Stope back	C	0	15	0.2

Factor C accounts for the influence of the stope wall orientation. A comparison of the geometry of the example mine with the sketches shown in Figures 14.4 and 14.5 suggests that the dominant modes of failure will be gravity falls from the stope back and buckling failure from the stope hangingwall and ends. The gravity adjustment factor is obtained from Figure 14.10 which gives $C = 2$, for the stope back, $C = 5.5$, for the hanging wall, and $C = 8$, for the stope ends.

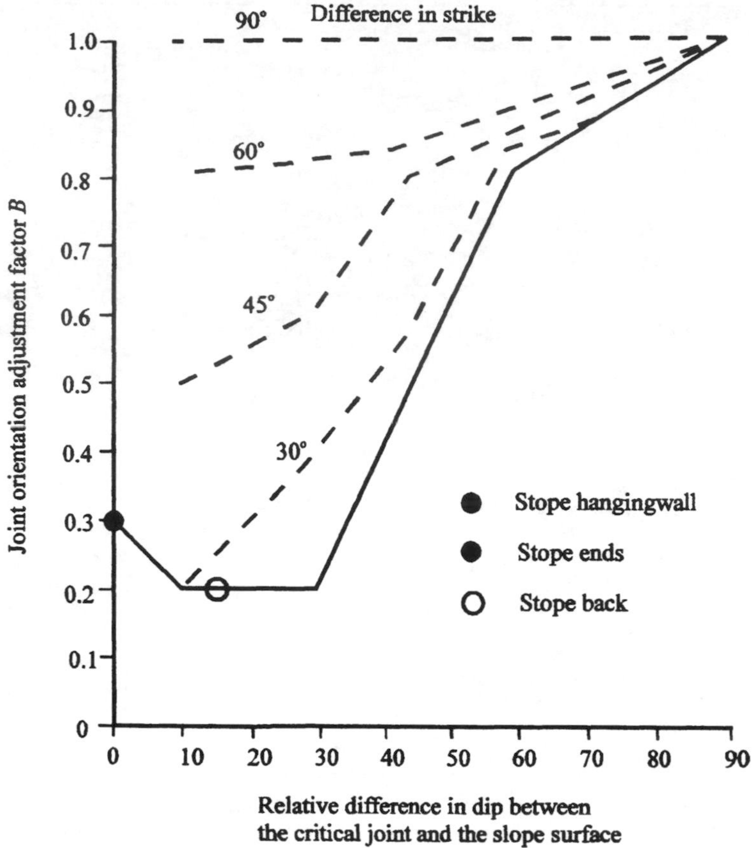

Figure 14.9: The correction factor B for the example mine.

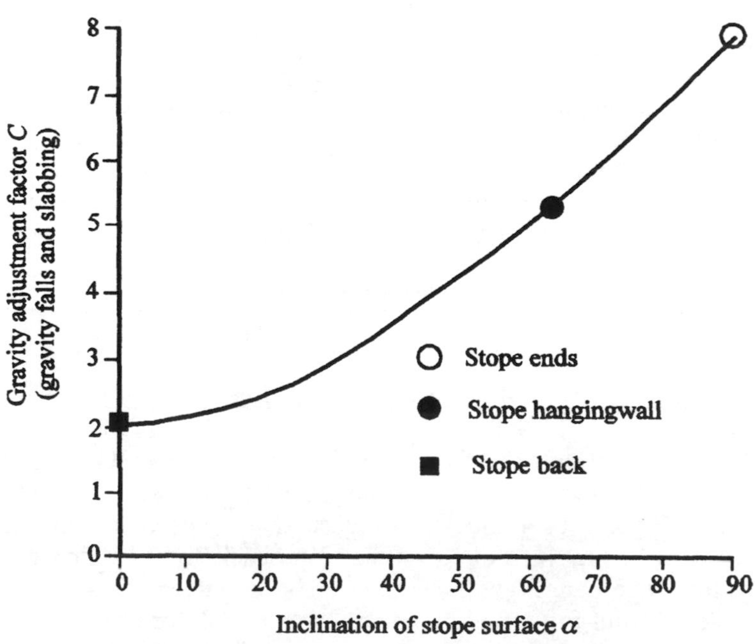

Figure 14.10: Gravity adjustment factors for example mine.

The stability number, N', for the stope back and hangingwall can now be calculated from Equation 14.1 and plotted on the stability graph as shown in Figure 14.11.

	Q'	A	B	C	N'
Stope back	6.3	0.25	0.2	2.0	0.63
Stope hangingwall	2.4	1.0	0.3	5.5	4.0

The stability graph gives the hydraulic radii of the stope that will be stable with and without support. The values of the hydraulic radii and associated stope widths are as follows:

	Back	Hangingwall
Known dimension	15 m span	27.6 m height
Hydraulic radius		
Stable	< 3	< 4.5
Unsupported transition	3 to 4.5	4.5 to 6.5
Stable with support	4.5 to 7.5	6.5 to 10
Supported transition	7.5 to 9	10 to 12
Calculated stope width		
Stable	< 10	< 13.4
Unsupported transition	10 to 22.5	13.4 to 24.6
Stable with support	22.5 to ∞	24.6 to 72.6
Supported transition		72.6 to 184

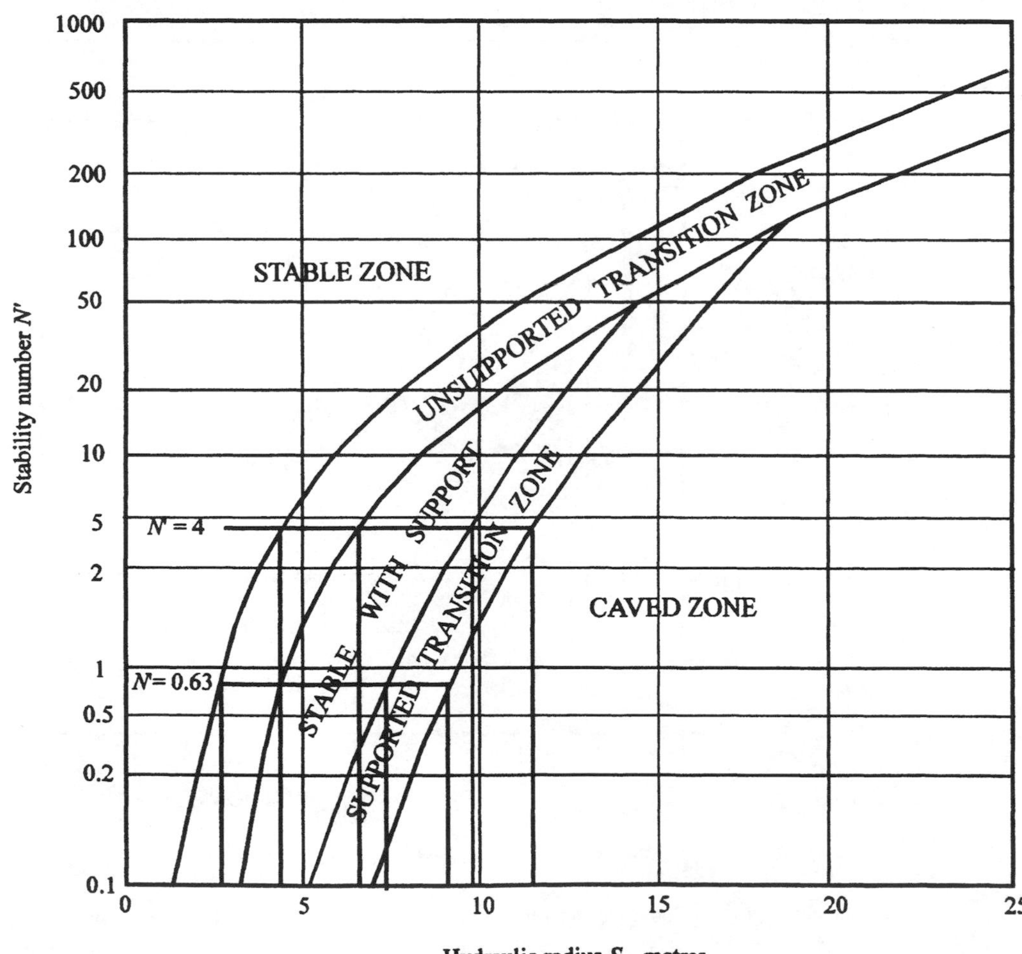

Figure 14.11: Stability of stope back and hangingwall for example mine.

The analysis indicates that the back is generally more critical than the hangingwall unless both are supported. For a 15 m span stope with a vertical height of 25 m, the width of the stope along strike should be less than 10 m for the stope to be stable without support. This strike distance is too short to allow for economical and safe development of drawpoints. Therefore, both the back and the hangingwall of the stope will have to be supported. The maximum safe strike length of a supported stope is controlled by the stability of the hanging wall and is about 75 m.

The decision on a reasonable strike length should be made on the basis of consideration of mining practicalities (overall orebody length, stope sequencing, drawpoint design etc.). If, for example, a reasonable strike length of the stope is determined to be 60 metres, a check calculation, using the same procedure as described above, will show that this stope is stable with support.

The preliminary design for the non-entry stope support can be carried out using Figure 14.7. The input data required for this analysis are given in the following table, and are plotted in Figure 14.12.

	S	$RQDxJ_n/S$	Bolt density (bolts/m$^{2)}$	Bolt spacing (m)
Back	6	1.1	.19–.33	1.7 to 2.3
Hangingwall	9.45	0.69	.23–.36	1.7 to 2.1
Ends	4.86	1.34	.16–.3	1.8 to 2.5

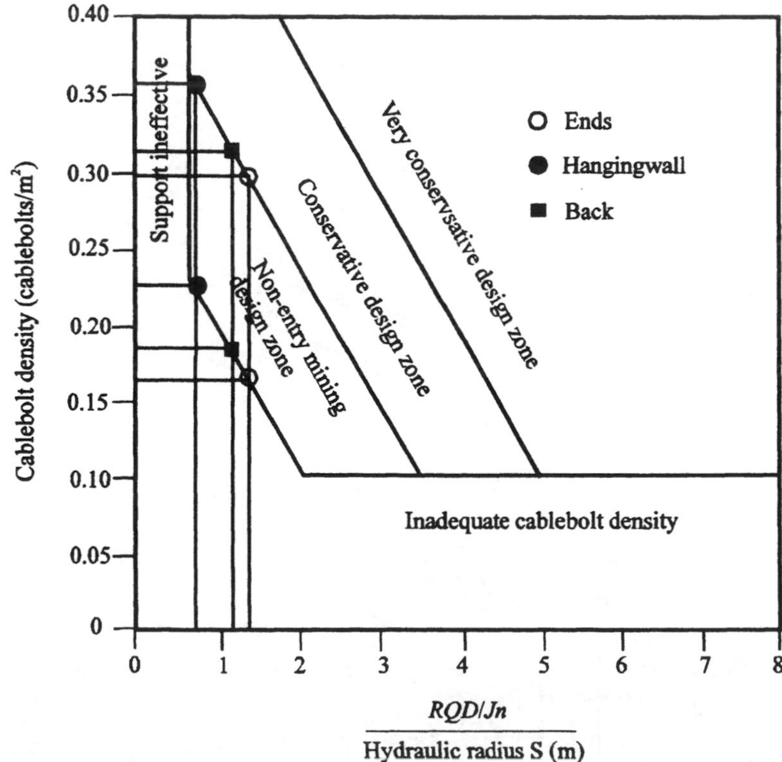

Figure 14.12: Cablebolt density for the preliminary design of the open stope at the example mine.

This analysis indicates that the cablebolts should be placed on regular patterns of about 2m x 2m spacing. In order to provide this denity of support, cablebolts will have to be installed from both the sublevels and the hangingwall cablebolt stub drifts, as indicated in the lower margin sketch on page 183.

Note that the hangingwall data plots close to the 'support ineffective' zone in Figure 14.12. As the design progresses, it would be advisable to find means of reducing the hydraulic radius of the stope, e.g., by early backfilling. If this can be achieved reliably during regular mining cycles, the cablebolt spacing may be increased to at least 2.5 m × 2.5 m, saving about one third of the cablebolts required without such a reduction.

Following the preliminary analysis, the design of the stope dimensions and support spacing should be refined as more information about the rock mass characteristics and operational constraints become available, and as the economics of mining the ore and the cost of the support are evaluated.

15 Shotcrete support

15.1 Introduction

The use of shotcrete for the support of underground excavations was pioneered by the civil engineering industry. Reviews of the development of shotcrete technology have been presented by Rose (1985), Morgan (1992) and Franzén (1992). Rabcewicz (1969) was largely responsible for the introduction of the use of shotcrete for tunnel support in the 1930s, and for the development of the New Austrian Tunnelling Method for excavating in weak ground.

In recent years the mining industry has become a major user of shotcrete for underground support. It can be expected to make its own contributions to this field as it has in other areas of underground support. The simultaneous working of multiple headings, difficulty of access and unusual loading conditions are some of the problems which are peculiar to underground mining and which require new and innovative applications of shotcrete technology.

An important area of shotcrete application in underground mining is in the support of 'permanent' openings such as ramps, haulages, shaft stations and crusher chambers. Rehabilitation of conventional rockbolt and mesh support can be very disruptive and expensive. Increasing numbers of these excavations are being shotcreted immediately after excavation. The incorporation of steel fibre reinforcement into the shotcrete is an important factor in this escalating use, since it minimises the labour intensive process of mesh installation.

Recent trials and observations suggest that shotcrete can provide effective support in mild rockburst conditions (McCreath and Kaiser, 1992, Langille and Burtney, 1992). While the results from these studies are still too limited to permit definite conclusions to be drawn, the indications are encouraging enough that more serious attention will probably be paid to this application in the future.

15.2 Shotcrete technology

Shotcrete is the generic name for cement, sand and fine aggregate concretes which are applied pneumatically and compacted dynamically under high velocity.

15.2.1 *Dry mix shotcrete*

As illustrated in Figure 15.1, the dry shotcrete components, which may be slightly pre-dampened to reduce dust, are fed into a hopper with continuous agitation. Compressed air is introduced through a rotating barrel or feed bowl to convey the materials in a continuous stream through the delivery hose. Water is added to the mix at the nozzle. Gunite, a proprietary name for dry-sprayed mortar used in the early 1900's, has fallen into disuse in favour of the more general term shotcrete.

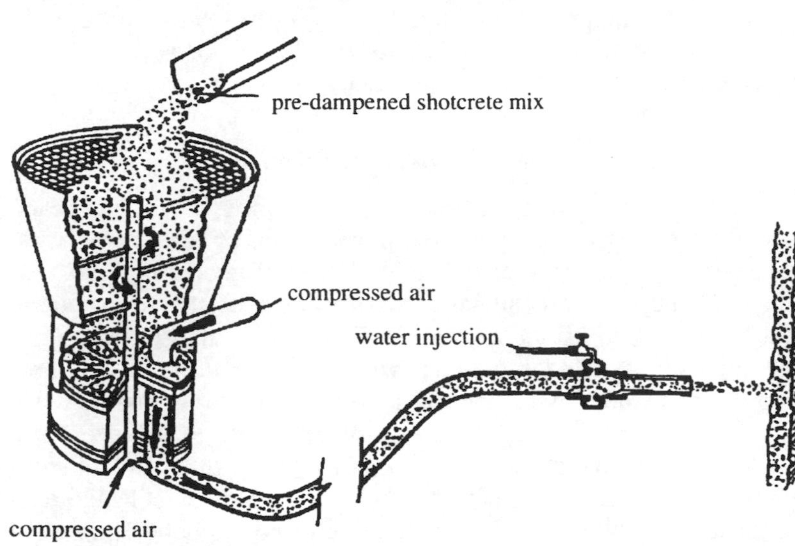

Figure 15.1: Simplified sketch of a typical dry mix shotcrete system. After Mahar et al. (1975).

Typical dry mix shotcrete machine.

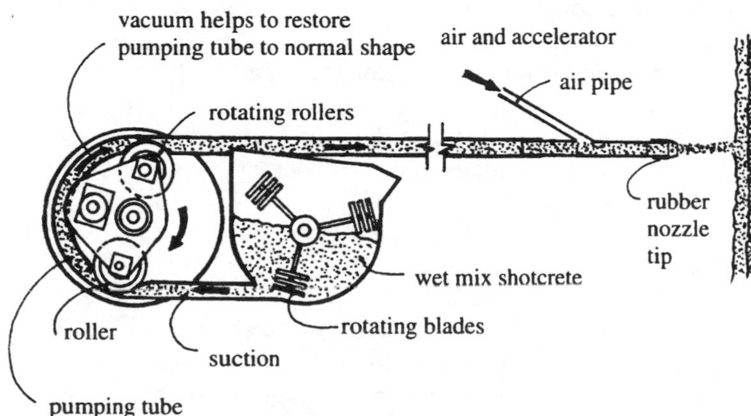

Figure 15.2: Typical wet mix shotcrete machine. After Mahar et al. (1975).

15.2.2 *Wet mix shotcrete*

In this case the shotcrete components and the water are mixed (usually in a truck mounted mixer) before delivery into a positive displacement pumping unit, which then delivers the mix hydraulically to the nozzle where air is added to project the material onto the rock surface.

The final product of either the dry or wet shotcrete process is very similar. The dry mix system tends to be more widely used in mining, because of inaccessibility for large transit mix trucks and because it generally uses smaller and more compact equipment. This can be moved around relatively easily in an underground mine environment. The wet mix system is ideal for high production applications, where a deep shaft or long tunnel is being driven and where access allows the application equipment and delivery trucks to operate on a more or less continuous basis. Decisions to use the dry or wet mix shotcrete process are usually made on a site-by-site basis.

15.2.3 *Steel fibre reinforced micro silica shotcrete*

Of the many developments in shotcrete technology in recent years, two of the most significant were the introduction of silica fume, used as a cementitious admixture, and steel fibre reinforcement.

Silica fume or micro silica is a by-product of the ferro silicon metal industry and is an extremely fine pozzolan. Pozzolans are cementitious materials which react with the calcium hydroxide produced during cement hydration. Silica fume, added in quantities of 8 to 13% by weight of cement, can allow shotcrete to achieve compressive strengths which are double or triple the value of plain shotcrete mixes. The result is an extremely strong, impermeable and durable shotcrete. Other benefits include reduced rebound, improved flexural strength, improved bond with the rock mass and the ability to place layers of up to 200 mm thick in a single pass because of the shotcrete's 'stickiness'. However, when using wet mix shotcrete, this stickiness decreases the workability of the material and superplaticizers are required to restore this workability.

Steel fibre reinforced shotcrete was introduced in the 1970s and has since gained world-wide acceptance as a replacement for traditional wire mesh reinforced plain shotcrete. The main role that reinforcement plays in shotcrete is to impart ductility to an otherwise brittle material. As pointed out earlier, rock support is only called upon to carry significant loads once the rock surrounding an underground excavation deforms. This means that unevenly distributed non-elastic deformations of significant magnitude may overload and lead to failure of the support system, unless that system has sufficient ductility to accommodate these deformations.

Typical steel fibre reinforced, silica fume shotcrete mix designs are summarised in Table 15.1. These mixes can be used as a starting point when embarking on a shotcrete programme, but it may be necessary to seek expert assistance to 'fine tune' the mix designs to suit site specific requirements. For many dry mix applications it may be advantageous to purchase pre-mixed shotcrete in bags of up to 1,500 kg capacity, as illustrated in Figure 15.3.

Table 15.1: Typical steel fibre reinforced silica fume shotcrete mix designs (After Wood, 1992).

Components	Dry mix		Wet mix	
	kg./m^3	% dry materials	kg./m^3	% wet materials
Cement	420	19.0	420	18.1
Silica fume additive	50	2.2	40	1.7
Blended aggregate	1,670	75.5	1,600	68.9
Steel fibres	60	2.7	60	2.6
Accelerator	13	0.6	13	0.6
Superplasticizer	-	-	6 litres	0.3
Water reducer	-	-	2 litres	0.1
Air entraining admixture	-	-	if required	
Water	controlled at nozzle		180	7.7
Total	2,213	100	2,321	100

Figure 15.3: Bagged pre-mixed dry shotcrete components being delivered into a hopper feeding a screw conveyor, fitted with a pre-dampener, which discharges into the hopper of a shotcrete machine.

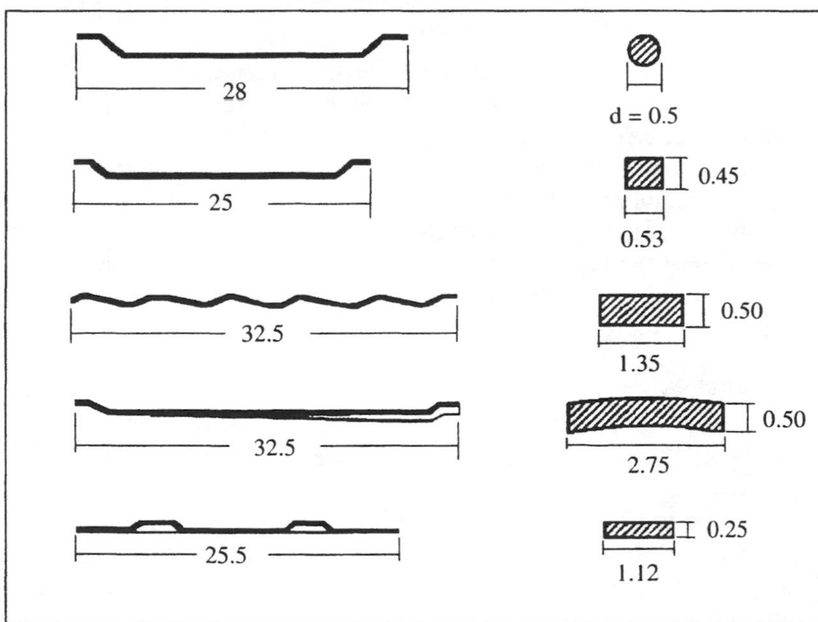

Figure 15.4. Steel fibre types available on the north American market. After Wood et al. (1993). (Note: all dimensions are in mm).

Figure 15.4 shows the steel fibre types which are currently available on the north American market. In addition to their use in shotcrete, these fibres are also widely used in concrete floor slabs for buildings, in airport runways and in similar concrete applications.

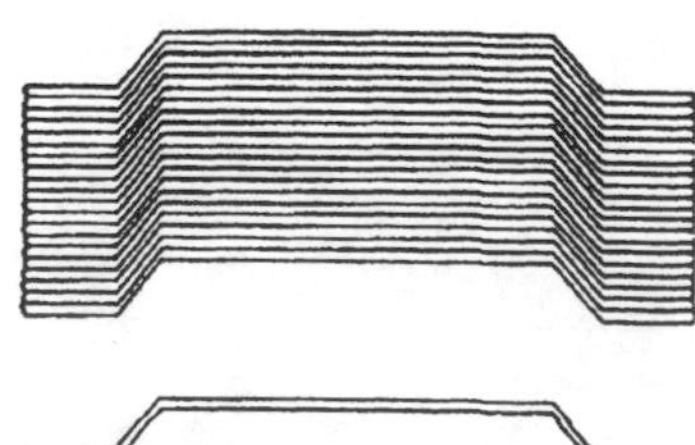

'Dramix' steel fibres used in slab bending tests by Kompen (1989). The fibres are glued together in bundles with a water soluble glue to facilitate handling and homogeneous distribution of the fibres in the shotcrete.

Wood et al. (1993) have reported the results of a comprehensive comparative study in which all of the fibres shown in Figure 15.4 were used to reinforce shotcrete samples, which were then subjected to a range of tests. Plain and fibre reinforced silica fume shotcrete samples were prepared by shooting onto vertical panels, using both wet and dry mix processes. The fibre reinforced samples all contained the same steel fibre dosage of 60 kg/m^3 (see Table 15.1). All the samples were cured under controlled relative humidity conditions and all were tested seven days after shooting.

These tests showed that the addition of steel fibres to silica fume shotcrete enhances both the compressive and flexural strength of the hardened shotcrete by up to 20%. A significant increase in ductility was also obtained in all the tests on fibre reinforced samples, compared with plain samples. While different fibres gave different degrees of improvement, all of the fibres tested were found to exceed the levels of performance commonly specified in north America (i.e. 7-day compressive strength of 30 MPa for dry mix, 25 MPa for wet mix and 7-day flexural strength of 4 MPa).

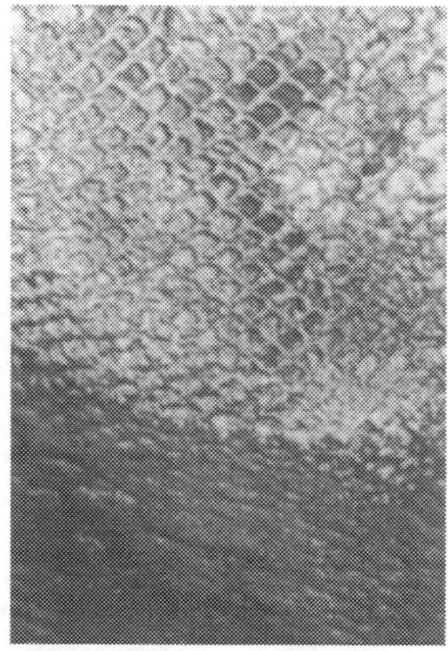

Chainlink mesh, while very strong and flexible, is not ideal for shotcrete application because it is difficult for the shotcrete to penetrate the mesh.

Kompen (1989) carried out bending tests on slabs of unreinforced shotcrete and shotcrete reinforced with 'Dramix'[1] steel fibres. The shotcrete had an unconfined compressive strength, determined from tests on cubes, of 50 MPa. The results of these tests are reproduced in Figure 15.5. The peak strength of these slabs increased by approximately 85% and 185% for 1.0 and 1.5 volume % of fibres, respectively. The ductility of the fibre reinforced slabs increased by approximately 20 and 30 times for the 1.0 and 1.5 volume % of fibres, respectively.

15.2.4 *Mesh reinforced shotcrete*

While steel fibre reinforced shotcrete has been widely accepted in both civil and mining engineering, mesh reinforced shotcrete is still widely used and is preferred in some applications. In very poor quality, loose rock masses, where adhesion of the shotcrete to the rock surface is poor, the mesh provides a significant amount of reinforcement, even without shotcrete. Therefore, when stabilising slopes in very poor quality rock masses or when building bulkheads for underground fill, weldmesh is frequently used to stabilise the surface or to provide reinforcement. In such cases, plain shotcrete is applied later to provide additional support and to protect the mesh against corrosion.

Welded wire mesh, firmly attached to the rock surface, provides excellent reinforcement for shotcrete.

Kirsten (1992, 1993) carried out a comprehensive set of laboratory bending tests on both mesh and fibre reinforced shotcrete slabs. The load versus deflection curves, which he obtained, were similar to those reported by Kompen, reproduced in Figure 15.5. He found that the load carrying capacity of the mesh and fibre reinforced shotcrete samples were not significantly different, but that the mesh reinforced samples were superior in bending with both point loads and uniformly distributed loads. He concluded that this was due to the more favourable location of the mesh reinforcement in the slabs subjected to bending.

[1] Manufactured by N.V. Bekaert S.A., B-8550 Zwevegem, Belgium.

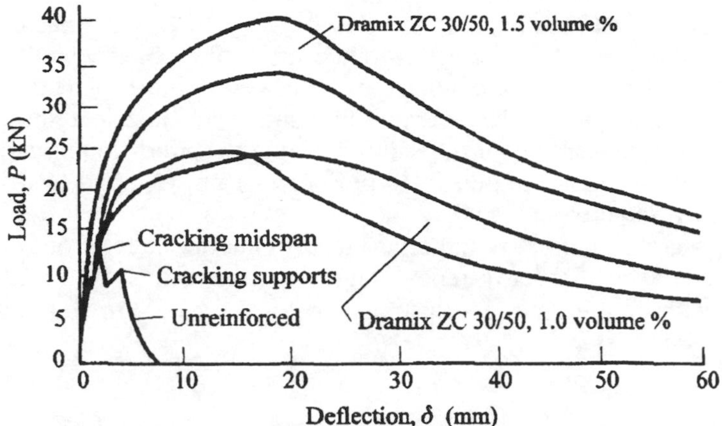

Figure 15.5: Load deflection curves for unreinforced and steel fibre reinforced shotcrete slabs tested in bending. After Kompen (1989).

Kirsten also concluded that the quality control, required to obtain a consistent dosage and uniform distribution of fibres in shotcrete, is more easily achieved in civil engineering than in mining applications. This is a reflection of the multiple working headings and the difficulties of access which are common problems associated with many mines. Under these circumstances, more reliable reinforcement will be obtained with mesh reinforced rather than fibre reinforced shotcrete. However, in large mines, in which many of the 'permanent' openings are similar to those on large civil engineering sites, these problems of quality control should not arise.

15.3 Shotcrete application

The quality of the final shotcrete product is closely related to the application procedures used. These procedures include: surface preparation, nozzling technique, lighting, ventilation, communications, and crew training.

Shotcrete should not be applied directly to a dry, dusty or frozen rock surface. The work area is usually sprayed with an air-water jet to remove loose rock and dust from the surface to be shot. The damp rock will create a good surface on which to bond the initial layer of shotcrete paste. The nozzleman commonly starts low on the wall and moves the nozzle in small circles working his way up towards the back, or roof. Care must be taken to avoid applying fresh materials on top of rebound or oversprayed shotcrete. It is essential that the air supply is consistent and has sufficient capacity to ensure the delivery of a steady stream of high velocity shotcrete to the rock face. Shooting distances are ideally about 1 to 1.5 metres. Holding the nozzle further from the rock face will result in a lower velocity flow of materials which leads to poor compaction and a higher proportion of rebound.

A well-trained operator can produce excellent quality shotcrete manually, when the work area is well-lit and well-ventilated, and when the crew members are in good communication with each other using prescribed hand signals or voice activated FM radio headsets. However, this is a very tiring and uncomfortable job, especially for overhead shooting, and compact robotic systems are increasingly being used to permit the operator to control the nozzle remotely. Typical robotic spray booms, used for shotcrete application in underground excavations, are illustrated in Figures 15.6, 15.7 and 15.8.

When shotcrete is applied to rock masses with well-defined water-bearing joints, it is important to provide drainage through the shotcrete layer in order to relieve high water pressures. Drain holes, fitted with plastic pipes as illustrated in Figure 15.9, are commonly

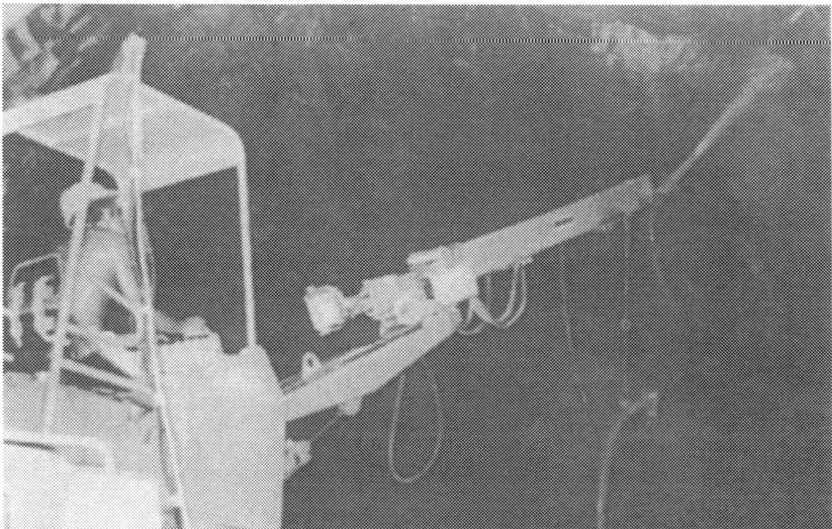

Figure 15.6: A truck mounted shotcrete robot being used in a large civil engineering tunnel. Note that the distance between the nozzle and the rock surface is approximately one metre.

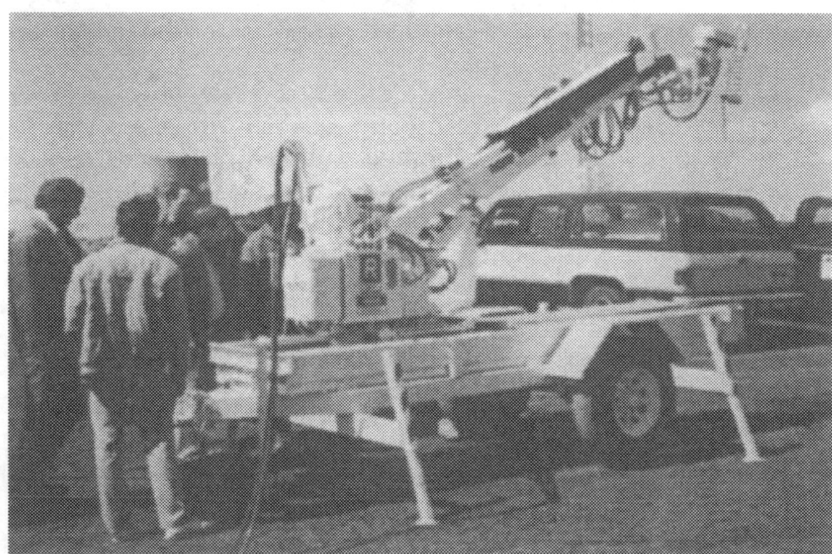

Figure 15.7: Compact trailer-mounted robot unit for remote controlled shotcrete application.

Figure 15.8: Shotcrete operator using a remotely controlled unit to apply shotcrete to a rock face in a large civil engineering excavation.

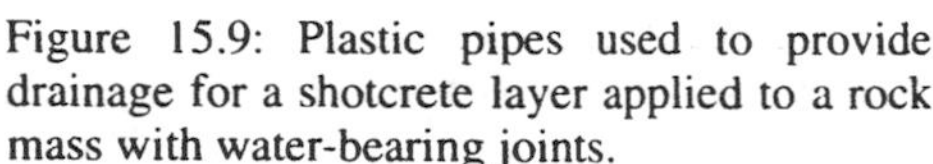

Figure 15.9: Plastic pipes used to provide drainage for a shotcrete layer applied to a rock mass with water-bearing joints.

used for this purpose. Where the water inflow is not restricted to a few specific features, a porous fibre mat can be attached to the rock surface before the shotcrete layer is applied. When practical to do so, the water from these drains should be collected and directed into a drainage ditch or sump.

15.4 Design of shotcrete support

The design of shotcrete support for underground excavations is a very imprecise process. However, one observation, which is commonly made by practical engineers with years of experience in using shotcrete underground, is that it almost always performs better than anticipated. There are many examples (very few of which are documented) where shotcrete has been used as a last act of desperation in an effort to stabilise the failing rock around a tunnel and, to most people's surprise, it has worked.

The complex interaction between the failing rock mass around an underground opening, and a layer of shotcrete of varying thickness with properties which change as it hardens, defies most attempts at theoretical analysis. The simplistic closed-form support-interaction analyses described in Chapter 9 give a very crude indication of the possible support action of shotcrete. It is only in recent years, with the development of powerful numerical tools such as the programs FLAC[2] and PHASES, that it has been possible to contemplate realistic analyses, which will explore the possible support-interaction behaviour of shotcrete. A clear understanding of shotcrete behaviour will require many more years of experience in the use of and in the interpretation of the results obtained from these programs. It is also important to recognise that shotcrete is very seldom used alone and its use in combination with rockbolts, cablebolts, lattice girders or steel sets further complicates the problem of analysing its contribution to support.

Current shotcrete support 'design' methodology relies very heavily upon rules of thumb and precedent experience. Wickham et al. (1972) related the thickness of a shotcrete tunnel lining to their Rock Structure Rating (*RSR*) and produced the plot given in Figure 4.2 in Chapter 4. Bieniawski (1989) gave recommendations on shotcrete thicknesses (in conjunction with rockbolts or steel sets) for different Rock Mass Ratings (*RMR*) for a 10 m span opening. These recommendations are summarised in Table 4.5 in Chapter 4. Grimstad and Barton (1993) have published an updated chart (reproduced in Figure 4.3 in Chapter 4) relating different support systems, including shotcrete and fibre reinforced shotcrete, to the Tunnelling Quality Index *Q*. Vandewalle (1990) collected various rules of thumb from a variety of sources and included them in his monograph.

Table 15.2 is a compilation of current shotcrete practice by the present authors, combining all of these empirical rules and adding in their own practical experience. The reader is warned, that this table can only be used as an approximate guide when deciding upon the

[2] Obtainable from ITASCA Consulting Group Inc.,Thresher Square East, 708 South Third Street, Suite 310, Minneapolis, Minnesota 55415, USA, Fax 1 612 371 4717

Table 15.2: Summary of recommended shotcrete applications in underground mining, for different rock mass conditions.

Rock mass description	Rock mass behaviour	Support requirements	Shotcrete application
Massive metamorphic or igneous rock. Low stress conditions.	No spalling, slabbing or failure.	None.	None.
Massive sedimentary rock. Low stress conditions.	Surfaces of some shales, siltstones, or claystones may slake as a result of moisture content change.	Sealing surface to prevent slaking.	Apply 25 mm thickness of plain shotcrete to permanent surfaces as soon as possible after excavation. Repair shotcrete damage due to blasting.
Massive rock with single wide fault or shear zone.	Fault gouge may be weak and erodible and may cause stability problems in adjacent jointed rock.	Provision of support and surface sealing in vicinity of weak fault of shear zone.	Remove weak material to a depth equal to width of fault or shear zone and grout rebar into adjacent sound rock. Weldmesh can be used if required to provide temporary rockfall support. Fill void with plain shotcrete. Extend steel fibre reinforced shotcrete laterally for at least width of gouge zone.
Massive metamorphic or igneous rock. High stress conditions.	Surface slabbing, spalling and possible rockburst damage.	Retention of broken rock and control of rock mass dilation.	Apply 50 mm shotcrete over weldmesh anchored behind bolt faceplates, or apply 50 mm of steel fibre reinforced shotcrete on rock and install rockbolts with faceplates; then apply second 25 mm shotcrete layer. Extend shotcrete application down sidewalls where required.
Massive sedimentary rock. High stress conditions.	Surface slabbing, spalling and possible squeezing in shales and soft rocks.	Retention of broken rock and control of squeezing.	Apply 75 mm layer of fibre reinforced shotcrete directly on clean rock. Rockbolts or dowels are also needed for additional support.
Metamorphic or igneous rock with a few widely spaced joints. Low stress conditions.	Potential for wedges or blocks to fall or slide due to gravity loading.	Provision of support in addition to that available from rockbolts or cables.	Apply 50 mm of steel fibre reinforced shotcrete to rock surfaces on which joint traces are exposed.
Sedimentary rock with a few widely spaced bedding planes and joints. Low stress conditions.	Potential for wedges or blocks to fall or slide due to gravity loading. Bedding plane exposures may deteriorate in time.	Provision of support in addition to that available from rockbolts or cables. Sealing of weak bedding plane exposures.	Apply 50 mm of steel fibre reinforced shotcrete on rock surface on which discontinuity traces are exposed, with particular attention to bedding plane traces.
Jointed metamorphic or igneous rock. High stress conditions.	Combined structural and stress controlled failures around opening boundary.	Retention of broken rock and control of rock mass dilation.	Apply 75 mm plain shotcrete over weldmesh anchored behind bolt faceplates or apply 75 mm of steel fibre reinforced shotcrete on rock, install rockbolts with faceplates and then apply second 25 mm shotcrete layer. Thicker shotcrete layers may be required at high stress concentrations.
Bedded and jointed weak sedimentary rock. High stress conditions.	Slabbing, spalling and possibly squeezing.	Control of rock mass failure and squeezing.	Apply 75 mm of steel fibre reinforced shotcrete to clean rock surfaces as soon as possible, install rockbolts, with faceplates, through shotcrete, apply second 75 mm shotcrete layer.
Highly jointed metamorphic or igneous rock. Low stress conditions.	Ravelling of small wedges and blocks defined by intersecting joints.	Prevention of progressive ravelling.	Apply 50 mm of steel fibre reinforced shotcrete on clean rock surface in roof of excavation. Rockbolts or dowels may be needed for additional support for large blocks.

Table 15.2: (continued).

Rock mass description	Rock mass behaviour	Support requirement	Shotcrete application
Highly jointed and bedded sedimentary rock. Low stress conditions.	Bed separation in wide span excavations and ravelling of bedding traces in inclined faces.	Control of bed separation and ravelling.	Rockbolts or dowels required to control bed separation. Apply 75 mm of fibre reinforced shotcrete to bedding plane traces before bolting.
Heavily jointed igneous or metamorphic rock, conglomerates or cemented rockfill. High stress conditions.	Squeezing and 'plastic' flow of rock mass around opening.	Control of rock mass failure and dilation.	Apply 100 mm of steel fibre reinforced shotcrete as soon as possible and install rockbolts, with face-plates, through shotcrete. Apply additional 50 mm of shotcrete if required. Extend support down sidewalls if necessary.
Heavily jointed sedimentary rock with clay coated surfaces. High stress conditions.	Squeezing and 'plastic' flow of rock mass around opening. Clay rich rocks may swell.	Control of rock mass failure and dilation.	Apply 50 mm of steel fibre reinforced shotcrete as soon as possible, install lattice girders or light steel sets, with invert struts where required, then more steel fibre reinforced shotcrete to cover sets or girders. Forepoling or spiling may be required to stabilise face ahead of excavation. Gaps may be left in final shotcrete to allow for movement resulting from squeezing or swelling. Gap should be closed once opening is stable.
Mild rockburst conditions in massive rock subjected to high stress conditions.	Spalling, slabbing and mild rockbursts.	Retention of broken rock and control of failure propagation.	Apply 50 to 100 mm of shotcrete over mesh or cable lacing which is firmly attached to the rock surface by means of yielding rockbolts or cablebolts.

type and thickness of shotcrete to be applied in a specific application. Modifications will almost certainly be required to deal with local variations in rock conditions and shotcrete quality.

Shotcrete cannot prevent deformation from taking place, especially in high stress environments. It can, however, assist in controlling deformation, particularly when used in combination with rockbolts, dowels or cables. Shotcrete support becomes very effective when bolt or cable installations are carried out after an initial shotcrete application. This allows the face plate loads to be transmitted over a large area to the underlying rock mass.

References

Anon. 1977. Description of rock masses for engineering purposes. Geological Society Engineering Group Working Party Report. *Q. J. Engng Geol.* **10**, 355-388.

Bajzelj, U., Likar, J., Zigman, F., Subelj, A. and Spek, S. 1992. Geotechnical analyses of the mining method using long cable bolts. In *Rock support in mining and underground construction, proc. int. symp. on rock support,* Sudbury, (eds. P.K. Kaiser and D.R. McCreath), 393-402. Rotterdam: Balkema.

Balmer, G. 1952. A general analytical solution for Mohr's envelope. *Am. Soc. Test. Mat.* **52**, 1260-1271.

Bandis, S.C. 1980. *Experimental studies of scale effects on shear strength, and deformation of rock joints.* Ph.D. thesis, University of Leeds.

Bandis, S.C. 1990. Mechanical properties of rock joints. In *Proc. Int. Soc. Rock Mech. symp. on rock joints,* Loen, Norway, (eds N. Barton and O. Stephansson), 125-140. Rotterdam: Balkema.

Barton, N.R. 1973. Review of a new shear strength criterion for rock joints. *Engng Geol.* **7**, 287-332.

Barton, N.R. 1974. *A review of the shear strength of filled discontinuities in rock.* Norwegian Geotech. Inst. Publ. No. 105. Oslo: Norwegian Geotech. Inst.

Barton, N.R. 1976. The shear strength of rock and rock joints. *Int. J. Mech. Min. Sci. & Geomech. Abstr.* **13**(10), 1-24.

Barton, N.R. 1987. *Predicting the behaviour of underground openings in rock. Manuel Rocha Memorial Lecture*, Lisbon. Oslo: Norwegian Geotech. Inst.

Barton, N.R. and Bandis, S.C. 1982. Effects of block size on the the shear behaviour of jointed rock. *23rd U.S. symp. on rock mechanics*, Berkeley, 739-760.

Barton, N.R. and Bandis, S.C. 1990. Review of predictive capabilites of JRC-JCS model in engineering practice. In *Rock joints, proc. int. symp. on rock joints,* Loen, Norway, (eds N. Barton and O. Stephansson), 603-610. Rotterdam: Balkema.

Barton, N.R. and Choubey, V. 1977. The shear strength of rock joints in theory and practice. *Rock Mech.* **10**(1-2), 1-54.

Barton, N.R., Lien, R. and Lunde, J. 1974. Engineering classification of rock masses for the design of tunnel support. *Rock Mech.* **6**(4), 189-239.

Barton, N., Løset, F., Lien, R. and Lunde, J. 1980. Application of the Q-system in design decisions. In *Subsurface space,* (ed. M. Bergman) **2,** 553-561. New York: Pergamon.

Barton, N., By, T.L., Chryssanthakis, L., Tunbridge, L., Kristiansen, J., Løset, F., Bhasin, R.K., Westerdahl, H. and Vik, G. 1992. Comparison of prediction and performance for a 62 m span sports hall in jointed gneiss. *Proc. 4th. int. rock mechanics and rock engineering conf.,* Torino. Paper 17.

Bieniawski, Z.T. 1967. Mechanism of brittle fracture of rock, parts I, II and III. *Int. J. Rock Mech. Min. Sci. & Geomech. Abstr.* **4**(4), 395-430.

Bieniawski, Z.T. 1973. Engineering classification of jointed rock masses. *Trans S. Afr. Inst. Civ. Engrs* **15**, 335-344.

Bieniawski, Z.T. 1974. Geomechanics classification of rock masses and its application in tunnelling. In *Advances in rock mechanics* **2** (A), 27-32. Washington, D.C.: Nat. Acad. Sci.

Bieniawski, Z.T. 1976. Rock mass classification in rock engineering. In *Exploration for rock engineering, proc. of the symp.,* (ed. Z.T. Bieniawski) **1**, 97-106. Cape Town: Balkema.

Bieniawski, Z.T. 1978. Determining rock mass deformability - experiences from case histories. *Int. J. Rock Mech. Min. Sci. & Geomech. Abstr.* **15**, 237-247.

Bieniawski, Z.T. 1979. The geomechanics classification in rock engineering applications. *Proc. 4th. congr., Int. Soc. Rock Mech.,* Montreux **2**, 41-48.

Bieniawski, Z.T. 1989. *Engineering rock mass classifications.* New York: Wiley.

Bouchier, F., Dib, E. and O'Flaherty, M. 1992. Practical improvements to installation of cable bolts: progress at Campbell Mine. In *Rock support in mining and underground construction, proc. int. symp. on rock support,* Sudbury, (eds P.K. Kaiser and D.R. McCreath), 311-318. Rotterdam: Balkema.

Brady, B.H.G. and Brown, E.T. 1985. *Rock mechanics for underground mining.* London: Allen and Unwin.

Brown, E.T. 1987. Introduction. *Analytical and computational methods in engineering rock mechanics,* (ed. E.T. Brown), 1-31. London: Allen and Unwin.

Brown, E.T. and Hoek, E. 1978. Trends in relationships between measured rock in situ stresses and depth. *Int. J. Rock Mech. Min. Sci. & Geomech. Abstr.* **15**, 211-215.

Brown, E.T., Bray, J.W., Ladanyi, B. and Hoek, E. 1983. Characteristic line calculations for rock tunnels. *J. Geotech. Engng Div., ASCE* **109**, 15-39.

Bywater, S. and Fuller, P.G. 1984. Cable support for lead open stope hanging walls at Mount Isa Mines Limited. In *Rock bolting: theory and application in mining and underground construction,* (ed. O. Stephansson), 539-556. Rotterdam: Balkema.

Carter, T.G. 1992. Prediction and uncertainties in geological engineering and rock mass characterization assessments. *Proc. 4th. int. rock mechanics and rock engineering conf.,* Torino. Paper 1.

Clegg, I.D. and Hanson, D.S. 1992. Ore pass design and support at Falconbridge Limited. In *Rock support in mining and underground construction, proc. int. symp. on rock support,* Sudbury, (eds P.K. Kaiser and D.R. McCreath), 219-225. Rotterdam: Balkema.

Clifford, R.L. 1974. Long rockbolt support at New Broken Hill Consolidated Limited. *Proc. Aus. Inst. Min. Metall.,* No. 251, 21-26.

Cook, N.G.W. (1965) The failure of rock. *Int. J. Rock Mech. Min. Sci. Geomech. Abstr.* **2**, 389-403.

Cording, E.J. and Deere, D.U. 1972. Rock tunnel supports and field measurements. *Proc. North American rapid excav. tunneling conf.,* Chicago, (eds. K.S. Lane and L.A. Garfield) **1**, 601-622. New York: Soc. Min. Engrs, Am. Inst. Min. Metall. Petrolm Engrs.

Crouch, S.L. and Starfield, A.M. 1983. *Boundary element methods in solid mechanics.* London: Allen and Unwin.

Cummings, R.A., Kendorski, F.S. and Bieniawski, Z.T. 1982. *Caving rock mass classification and support estimation.* U.S. Bureau of Mines Contract Report #J0100103. Chicago: Engineers International Inc.

Cundall, P.A. 1971. A computer model for simulating progressive large scale movements in blocky rock systems. In *Rock fracture, proc. int. symp. on rock fracture,* Nancy, Paper 2-8.

Davis, W.L. 1977. Initiation of cablebolting at West Coast Mines, Rosebury. *Proc. Aust. Inst. Min. Metall. conf.,* Tasmania, 215-225.

Deere, D.U. 1989. *Rock quality designation (RQD) after 20 years.* U.S. Army Corps Engrs Contract Report GL-89-1. Vicksburg, MS: Waterways Experimental Station.

Deere, D.U. and Deere, D.W. 1988. The rock quality designation (RQD) index in practice. In *Rock classification systems for engineering purposes,* (ed. L. Kirkaldie), ASTM Special Publication 984, 91-101. Philadelphia: Am. Soc. Test. Mat.

Deere, D.U. and Miller, R.P. 1966. *Engineering classification and index properties of rock.* Technical Report No. AFNL-TR-65-116. Albuquerque, NM: Air Force Weapons Laboratory.

Deere, D.U., Hendron, A.J., Patton, F.D. and Cording, E.J. 1967. Design of surface and near surface construction in rock. In *Failure and breakage of rock, proc. 8th U.S. symp. rock mech.,* (ed. C. Fairhurst), 237-302. New York: Soc. Min. Engrs, Am. Inst. Min. Metall. Petrolm Engrs.

Diederichs, M.S., Pieterse, E., Nosé, J. and Kaiser, P.K. 1993. A model for evaluating cable bond strength: an update. *Proc. Eurock '93,* Lisbon. (in press).

Dorsten, V., Frederick, F.H. and Preston, H.K. 1984. Epoxy coated seven-wire strand for prestressed concrete. *Prestressed Concrete Inst. J.* **29**(4), 1-11.

Doruk, P. 1991. *Analysis of the laboratory strength data using the original and modified Hoek-Brown failure criteria.* MASc thesis, Dept. Civil Engineering, University of Toronto.

Duncan Fama, M.E. 1993. Numerical modelling of yield zones in weak rocks. In *Comprehensive rock engineering,* (ed. J.A. Hudson) **2**, 49-75. Oxford: Pergamon.

Engelder, T. and Sbar, M.L. 1984. Near-surface in situ stress: introduction. *J. Geophys. Res.* **89**, 9321-9322. Princeton, NJ: Princeton University Press.

Ewy, R.T. and Cook, N.G.W. 1990. Deformation and fracture around cylindrical openings in rock. Parts I & II. *Int. J. Rock Mech. Min. Sci. Geomech. Abstr.* **27**, 387-427.

Fabjanczyk, M.W. 1982. Review of ground support practice in Australian underground metalliferous mines. *Proc. Aus. Inst. Min. Metall. conf.*, Melbourne, 337-349. Melbourne: Aust. Inst. Min. Metall.

Fairhurst, C. and Cook, N.G.W. 1966. The phenomenon of rock splitting parallel to a free surface under compressive stress. *Proc. 1st. congr. Int. Soc. Rock Mech.*, Lisbon **1**, 687-692.

Fenner, R. 1938. Untersuchungen zur Erkenntnis des Gebirgsdruckes. *Glukauf* **74**, 681-695, 705-715.

Franklin, J.A. and Hoek, E. 1970. Developments in triaxial testing equipment. *Rock Mech.* **2**, 223-228. Berlin: Springer-Verlag.

Franzén, T. 1992. Shotcrete for underground support - a state of the art report with focus on steel fibre reinforcement. In *Rock support in mining and underground construction, proc. int. symp. rock support,* Sudbury, (eds P.K. Kaiser and D.R. McCreath), 91-104. Rotterdam: Balkema.

Fuller, P.G. 1981. Pre-reinforcement of cut and fill stopes. In *Application of rock mechanics to cut and fill mining,* (eds O. Stephansson and M.J. Jones), 55-63. London: Instn Min. Metall.

Fuller, P.G. 1984. Cable support in mining - a keynote lecture. In *Rock bolting: theory and application in mining and underground construction,* (ed. O. Stephansson), 511-522. Rotterdam: Balkema.

Garford Pty Ltd. 1990. *An improved, economical method for rock stabilisation*. 4p. Perth.

Griffith, A.A. 1921. The phenomenon of rupture and flow in solids. *Phil. Trans. Roy. Soc.*, London **A221**, 163-198.

Griffith, A.A. 1924. Theory of rupture. *Proc. 1st congr. applied mechanics*, Delft, 55-63. Delft: Technische Bockhandel en Drukkerij.

Grimstad, E. and Barton, N. 1993. Updating the Q-System for NMT. *Proc. int. symp. on sprayed concrete - modern use of wet mix sprayed concrete for underground support*, Fagernes, (eds Kompen, Opsahl and Berg). Oslo: Norwegian Concrete Assn.

Harr, M.E. 1987. *Reliability-based design in civil engineering*. New York: McGraw-Hill.

Hatzor, Y. and Goodman, R.E. 1992. Application of block theory and the critical key block concept in tunneling; two case histories. In *Proc. Int. Soc. Rock Mech. conf. on fractured and jointed rock masses,* Lake Tahoe, California, 632-639.

Herget, G. 1988. *Stresses in rock.* Rotterdam: Balkema.

Hoek, E. 1965. *Rock fracture under static stress conditions*. Ph.D. thesis, University of Cape Town.

Hoek, E. 1983. Strength of jointed rock masses, 23rd. Rankine Lecture. *Géotechnique* **33**(3), 187-223.

Hoek, E. 1989. A limit equilibrium analysis of surface crown pillar stability. In *Surface crown pillar evaluation for active and abandoned metal mines,* (ed. M.C. Betourney), 3-13. Ottawa: Dept. Energy, Mines & Resources Canada.

Hoek, E. and Bray, J.W. 1981. *Rock slope engineering*. 3rd edn. London: Instn Min. Metall.

Hoek, E., and Brown, E.T. 1980a. *Underground excavations in rock.* London: Instn Min. Metall.

Hoek, E. and Brown, E.T. 1980b. Empirical strength criterion for rock masses. *J. Geotech. Engng Div., ASCE* **106**(GT9), 1013-1035.

Hoek, E. and Brown, E.T. 1988. The Hoek-Brown failure criterion - a 1988 update. In *Rock engineering for underground excavations, proc. 15th Canadian rock mech. symp.,* (ed. J.C. Curran), 31-38. Toronto: Dept. Civ. Engineering, University of Toronto.

Hoek, E. and Moy, D. 1993. Design of large powerhouse caverns in weak rock. In *Comprehensive rock engineering,* (ed. J.A. Hudson) **5**, 85-110. Oxford: Pergamon.

Hoek, E., Wood, D. and Shah, S. 1992. A modified Hoek-Brown criterion for jointed rock masses. *Proc. rock characterization, symp. Int. Soc. Rock Mech.: Eurock '92*, (ed. J.A. Hudson), 209-214. London: Brit. Geol. Soc.

Hunt, R.E.B. and Askew, J.E. 1977. Installation and design guidelines for cable dowel ground support at ZC/NBHC. *Proc. underground operators conf.*, Broken Hill, 113-122. Melbourne: Aus. Inst. Min. Metall.

Hutchins, W.R., Bywater, S., Thompson, A.G. and Windson, C.R. 1990. A versatile grouted cable dowel reinforcing system for rock. *Proc. Aus. Inst. Min. Metall.* **1**, 25-29.

Hyett, A.J., Bawden, W.F. and Coulson, A.L. 1992. Physical and mechanical properties of normal Portland cement pertaining to fully grouted cable bolts. In *Rock support in mining and underground construction, proc. int. symp. rock support,* Sudbury, (eds. P.K. Kaiser and D.R. McCreath), 341-348. Rotterdam: Balkema.

Hyett, A.J., Bawden, W.F. and Reichert, R.D. 1992. The effect of rock mass confinement on the bond strength of fully grouted cable bolts. *Int. J. Rock Mech. Min. Sci. & Geomech. Abstr.* **29**(5), 503-524.

Hyett, A.J., Bawden, W.F., Powers, R. and Rocque, P. 1993. The nutcase cable. In *Innovative mine design for the 21st century,* (eds W.F. Bawden and J.F. Archibald), 409-419. Rotterdam: Balkema.

Iman, R.L., Davenport, J.M. and Zeigler, D.K. 1980. *Latin Hypercube sampling. (a program user's guide).* Technical Report SAND 79-1473. Albuquerque, NM: Sandia Laboratories.

International Society for Rock Mechanics. 1981. *Rock characterization, testing and monitoring - ISRM suggested methods.* Oxford: Pergamon.

International Society for Rock Mechanics Commission on Standardisation of Laboratory and Field Tests. 1978. Suggested methods for the quantitative description of discontinuities in rock masses. *Int. J. Rock Mech. Min. Sci. & Geomech. Abstr.* **15**, 319-368.

Jaeger, J.C. 1971. Friction of rocks and stability of rock slopes. The 11th Rankine Lecture. *Géotechnique* **21**(2), 97-134.

Jirovec. P. 1978. Wechselwirkung zwischen anker und gebirge. *Rock Mech.* Suppl. 7, 139-155.

Kaiser, P.K., Hoek, E. and Bawden, W.F. 1990. A new initiative in Canadian rock mechanics research. *Proc. 31st US rock mech. symp.*, Denver, 11-14.

Kaiser, P.K., Yazici, S. and Nosé, J. 1992. Effect of stress change on the bond strength of fully grouted cables. *Int. J. Rock Mech.. Min. Sci. Geomech. Abstr.* **29**(3), 293-306.

Kemeny, J.M. and Cook, N.G.W. 1987. Crack models for the failure of rock under compression. In *Proc. 2nd int. conf. on constitutive laws for engineering materials, theory and applications*, (eds C.S. Desai, E. Krempl, P.D. Kiousis and T. Kundu) 1, 879-887. Tucson, AZ: Elsevier.

Kendorski, F., Cummings, R., Bieniawski, Z.T. and Skinner, E. 1983. Rock mass classification for block caving mine drift support. *Proc. 5th congr. Int. Soc. Rock Mech.*, Melbourne, B51-B63. Rotterdam: Balkema.

Kirsch, G. 1898. Die Theorie der Elastizität und die Bedürfnisse der festigkeitslehre. *Veit. Deut. Ing.*, **42**(28), 797-807.

Kirsten, H.A.D. 1992. Comparative efficiency and ultimate strength of mesh- and fibre-reinforced shotcrete as determined from full-scale bending tests. *J. S. Afr. Inst. Min. Metall.* Nov., 303-322.

Kirsten, H.A.D. 1993. Equivalence of mesh- and fibre-reinforced shotcrete at large deflections. *Can. Geotech. J.* **30**, 418-440.

Kompen, R. 1989. Wet process steel fibre reinforced shotcrete for rock support and fire protection, Norwegian practice and experience. In *Proc. underground city conf.*, Munich, (ed. D. Morfeldt), 228-237.

Lang, T.A. 1961. Theory and practice of rockbolting. *Trans Amer. Inst. Min. Engrs* **220**, 333-348.

Langille, C.C. and Burtney, M.W. 1992. Effectiveness of shotcrete and mesh support in low energy rockburst conditions at INCO's Creighton mine. In *Rock support in mining and underground construction, proc. int. symp. rock support,* Sudbury, (eds. P.K. Kaiser and D.R. McCreath), 633-638. Rotterdam: Balkema.

Lappalainen, P., Pulkkinen, J. and Kuparinen, J. 1984. Use of steel strands in cable bolting and rock bolting. In *Rock bolting: theory and application in mining and underground construction*, (ed. O. Stephansson), 557-562. Rotterdam: Balkema.

Lajtai, E.Z. 1982. *The fracture of Lac du Bonnet granite*. Contract Report. Pinawa, Ontario: Atomic Energy of Canada.

Lajtai, E.Z. and Lajtai, V.N. 1975. The collapse of cavities. *Int. J. Rock Mech. Min. Sci. Geomech. Abstr.* **12,** 81-86.

Lau, J.S.O. and Gorski, B. 1991. The post failure behaviour of Lac du Bonnet red granite. CANMET Divisional Report MRL 91-079(TR). Ottawa: Dept. Energy Mines Resources.

Laubscher, D.H. 1977. Geomechanics classification of jointed rock masses - mining applications. *Trans. Instn. Min. Metall.* **86**, A1-8.

Laubscher, D.H. 1984. Design aspects and effectiveness of support systems in different mining conditions. *Trans Instn. Min. Metall.* **93**, A70 - A82.

Laubscher, D.M. and Page, C.H. 1990. The design of rock support in high stress or weak rock environments. *Proc. 92nd Can. Inst. Min. Metall. AGM*, Ottawa, Paper # 91.

Laubscher, D.H. and Taylor, H.W. 1976. The importance of geomechanics classification of jointed rock masses in mining operations. In *Exploration for rock engineering*, (ed. Z.T. Bieniawski) **1**, 119-128. Cape Town: Balkema.

Lauffer, H. 1958. Gebirgsklassifizierung für den Stollenbau. *Geol. Bauwesen* **24**(1), 46-51.

Lorig, L.J. and Brady, B.H.G. 1984. A hybrid computational scheme for excavation and support design in jointed rock media. In *Design and performance of underground excavations*, (eds E.T. Brown and J.A. Hudson), 105-112. London: Brit. Geotech. Soc.

Løset, F. 1992. Support needs compared at the Svartisen Road Tunnel. *Tunnels and Tunnelling*, June.

Love, A.E.H. 1927. *A treatise on the mathematical theory of elasticity*. New York: Dover.

Mahar, J.W., Parker, H.W. and Wuellner, W.W. 1975. *Shotcrete practice in underground construction*. US Dept. Transportation Report FRA-OR&D 75-90. Springfield, VA: Nat. Tech. Info. Service.

Marshall, D. 1963. Hangingwall control at Willroy. *Can. Min. Metall. Bull.* **56**, 327-331.

Martin, C.D. 1990. Characterizing in situ stress domains at the AECL Underground Research Laboratory. *Can. Geotech. J.* **27**, 631-646.

Martin, C.D. 1993. *The strength of massive Lac du Bonnet granite around underground openings*. Ph.D. thesis, Winnipeg, Manitoba: Dept. Civil Engineering, University of Manitoba.

Martin, C.D. and Simmons, G.R. 1992. The Underground Research Laboratory, an opportunity for basic rock mechanics. *Int. Soc. Rock Mech. News J.* **1**(1)5-12.

Mathews, K.E. and Edwards, D.B. 1969. Rock mechanics practice at Mount Isa Mines Limited, Australia. *Proc. 9th Commonwealth min. metall. congr.*, Paper 32. London: Instn Min. Metall.

Mathews, K.E., Hoek, E., Wyllie, D.C. and Stewart, S.B.V. 1981. *Prediction of stable excavations for mining at depth below 1000 metres in hard rock.* CANMET Report DSS Serial No. OSQ80-00081, DSS File No. 17SQ.23440-0-9020. Ottawa: Dept. Energy, Mines and Resources.

Matthews, S.M., Thompson, A.G., Windsor, C.R. and O'Bryan, P.R. 1986. A novel reinforcing system for large rock caverns in blocky rock masses. In *Large rock caverns*, (ed. K.H.O. Saari) **2**, 1541-1552. Oxford: Pergamon.

Mathews, S.M., Tillman, V.H. and Worotnicki, G. 1983. A modified cablebolt system for support of underground openings. *Proc. Aust. Inst. Min. Metall. annual conf.*, Broken Hill. 243-255.

McCreath, D.R. and Kaiser, P.K. 1992. Evaluation of current support practices in burst-prone ground and preliminary guidelines for Canadian hardrock mines. In *Rock support in mining and underground construction, proc. int. symp. rock support*, Sudbury, (eds P.K. Kaiser and D.R. McCreath), 611-619. Rotterdam: Balkema.

Merritt, A.H. 1972. Geologic prediction for underground excavations. *Proc. North American. rapid excav. tunneling conf.*, Chicago, (eds K.S. Lane and L.A. Garfield) **1,** 115-132. New York: Soc. Min. Engrs, Am. Inst. Min. Metall. Petrolm Engrs.

Morgan, D.R. 1993. Advances in shotcrete technology for support of underground openings in Canada.In *Shotcrete for underground support V, proc. engineering foundation conf.*, Uppsala, (eds J.C. Sharp and T. Franzen), 358-382. New York: Am. Soc. Civ. Engrs.

Muskhelishvili, N.I. 1953. *Some basic problems of the mathematical theory of elasticity*. 4th edn, translated by J.R.M. Radok. Gronigen: Noordhoff.

Nguyen, V. U. and Chowdhury, R.N. 1985. Simulation for risk analysis. *Geotechnique* **35**(1), 47-58.

Nickson, S.D. 1992. *Cable support guidelines for underground hard rock mine operations*. MASc. thesis, Dept. Mining and Mineral Processing, University of British Columbia.

Ortlepp, D.W. 1992. The design of the containment of rockburst damage in tunnels - an engineering approach. In *Rock support in mining and underground construction, proc. int. symp. on rock support,* Sudbury, (eds P.K. Kaiser and D.R. McCreath), 593-609. Rotterdam: Balkema.

Ortlepp, W. D.1993. Invited lecture: The design of support for the containment of rockburst damage in tunnels - an engineering approach. In *Rock support in mining and underground construction, proc. int. symp. on rock support*, Sudbury,(eds P.K. Kaiser and D.R. McCreath), 593-609. Rotterdam: Balkema.

Ortlepp, W. D. and Gay, N.C. 1984. Performance of an experimental tunnel subjected to stresses ranging form 50 MPa to 230 MPa. In *Design and performance of underground excavations*, (eds E.T. Brown and J.A. Hudson), 337-346. London: Brit. Geotech. Soc.

Otter, J.R.H., Cassell, A.C. and Hobbs, R.E. 1966. Dynamic relaxation. *Proc. Instn Civ. Engrs* **35**, 633-665.

Pacher, F., Rabcewicz, L. and Golser, J. 1974. Zum der seitigen Stand der Gebirgsklassifizierung in Stollen-und Tunnelbau. *Proc. XXII Geomech. colloq.*, Salzburg, 51-58.

Palmström, A. 1982. The volumetric joint count - a useful and simple measure of the degree of rock jointing. *Proc. 4th congr. Int. Assn Engng Geol.*, Delhi **5**, 221-228.

Patton, F.D. 1966. Multiple modes of shear failure in rock. *Proc. 1st congr. Int. Soc. Rock Mech.*, Lisbon **1**, 509-513.

Pelli, F., Kaiser, P.K. & Morgenstern, N.R. 1991. An interpretation of ground movements recorded during construction of the Donkin-Morien tunnel. *Can. Geotech. J.* **28**(2), 239-254

Pine, R.J. 1992. Risk analysis design applications in mining geomechanics. *Trans. Instn Min. Metall.* **101**, A149-158.

Potvin, Y. 1988. *Empirical open stope design in Canada*. Ph.D. thesis, Dept. Mining and Mineral Processing, University of British Columbia.

Potvin, Y. and Milne, D. 1992. Empirical cable bolt support design. In *Rock Support in mining and underground construction, proc. int. symp. on rock support*, Sudbury, (eds P.K. Kaiser and D.R. McCreath), 269-275. Rotterdam: Balkema.

Potvin, Y., Hudyma, M.R. and Miller, H.D.S. 1989. Design guidelines for open stope support. *Bull. Can. Min. Metall.* **82**(926), 53-62.

Rabcewicz, L.V. 1969. Stability of tunnels under rockload. *Water Power* **21**(6-8), 225-229, 266-273, 297-304.

Ritter, W. 1879. *Die Statik der Tunnelgewölbe*. Berlin: Springer.

Rose, D. 1985. Steel fibre reinforced shotcrete for tunnel linings: the state of the art. *Proc. North American rapid excav. tunneling conf.* **1**, 392-412. New York: Soc. Min. Engrs, Am. Inst. Min. Metall. Petrolm Engrs.

Rosenbleuth, E. 1981. Two-point estimates in probabilities. *J. Appl. Math. Modelling* **5**, October, 329-335.

Salamon, M.D.G. 1974. Rock mechanics of underground excavations. In *Advances in rock mechanics, proc. 3rd congr. Int. Soc. Rock Mech.*, Denver **1B**, 951-1009. Washington, D.C.: Nat. Acad. Sci.

Savin, G.N. 1961. *Stress concentrations around holes*. London: Pergamon.

Schmuck, C.H. 1979. Cable bolting at the Homestake gold mine. *Mining Engineering*, December, 1677-1681.

Scott, J.J. 1976. Friction rock stabilizers - a new rock reinforcement method. In *Monograph on rock mechanics applications in mining*, (eds W.S. Brown, S.J. Green and W.A. Hustrulid), 242-249. New York: Soc. Min. Engrs, Am. Inst. Min. Metall. Petrolm Engrs.

Scott, J.J. 1983. Friction rock stabilizer impact upon anchor design and ground control practices. In *Rock bolting: theory and application in underground construction*, (ed. O. Stephansson), 407-418. Rotterdam: Balkema.

Serafim, J.L. and Pereira, J.P. 1983. Consideration of the geomechanical classification of Bieniawski. *Proc. int. symp. on engineering geology and underground construction*, Lisbon **1**(II), 33-44.

Shah, S. 1992. *A study of the behaviour of jointed rock masses.* Ph.D. thesis, Dept. Civil Engineering, University of Toronto.

Sheory, P.R. 1994. A theory for in situ stresses in isotropic and transversely isotropic rock. *Int. J. Rock Mech. Min. Sci. & Geomech. Abstr.* **31**(1), 23-34.

Stillborg, B. 1994. *Professional users handbook for rock bolting,* 2nd edn. Clausthal- Zellerfeld: Trans Tech Publications.

Startzman, R.A. and Wattenbarger, R.A. 1985. An improved computation procedure for risk analysis problems with unusual probability functions. *Proc. symp. Soc. Petrolm Engrs hydrocarbon economics and evaluation,* Dallas.

Terzaghi, K. 1925. Erdbaumechanik auf Bodenphysikalischer Grundlage. Vienna: Franz Deuticke.

Terzaghi, K. 1946. Rock defects and loads on tunnel supports. In *Rock tunneling with steel supports,* (eds R. V. Proctor and T. L. White) **1**, 17-99. Youngstown, OH: Commercial Shearing and Stamping Company.

Terzaghi, K. and Richart, F.E. 1952. Stresses in rock about cavities. *Geotechnique* **3**, 57-90.

Thorn, L.J. and Muller, D.S. 1964. Prestressed roof support in underground engine chambers at Free State Geduld Mines Ltd. *Trans. Assn Mine Mngrs S. Afr.*, 411-428.

Tyler, D.B., Trueman, R.T. and Pine, R.J. 1991. Rockbolt support design using a probabilistic method of key block analysis. In *Rock mechanics as a multidisciplinary science*, (ed. J.C. Roegiers), 1037-1047. Rotterdam: Balkema.

VSL Systems Ltd. 1982. Slab post tensioning. 12p. Switzerland.

Vandewalle, M. 1993. *Dramix: Tunnelling the world.* 3rd edn. Zwevegem, Belgium: N.V. Bekaert S.A.

von Kimmelmann, M.R., Hyde, B. and Madgwick, R.J. 1984. The use of computer applications at BCL Limited in planning pillar extraction and the design of mine layouts. In *Design and performance of underground excavations*, (eds E.T. Brown and J.A. Hudson), 53-64. London: Brit. Geotech. Soc.

Whitman, R.V. 1984. Evaluating calculated risk in geotechnical engineering. *J. Geotech. Engng, ASCE,* **110**(2), 145-186.

Wickham, G.E., Tiedemann, H.R. and Skinner, E.H. 1972. Support determination based on geologic predictions. In *Proc. North American rapid excav. tunneling conf.*, Chicago, (eds K.S. Lane and L.A. Garfield), 43-64. New York: Soc. Min. Engrs, Am. Inst. Min. Metall. Petrolm Engrs.

Windsor, C.R. 1990. *Ferruled strand.* Unpublished memorandum. Perth: CSIRO.

Windsor, C.R. 1992. Cable bolting for underground and surface excavations. In *Rock support in mining and underground construction, proc. int. symp. on rock support*, Sudbury, (eds P.K. Kaiser and D.R. McCreath), 349-376. Rotterdam: Balkema.

Wood, D.F. 1992. Specification and application of fibre reinforced shotcrete. In *Rock support in mining and underground construction, proc. int. symp. on rock support*, Sudbury, (eds. P.K. Kaiser and D.R. McCreath), 149-156. Rotterdam: Balkema.

Wood, D.F., Banthia, N. and Trottier, J-F. 1993. A comparative study of different steel fibres in shotcrete. *In Shotcrete for underground support VI,* Niagara Falls, 57-66. New York: Am. Soc. Civ. Engrs.

Yazici, S. and Kaiser, P.K. 1992. Bond strength of grouted cable bolts. *Int J. Rock Mech. Min. Sci. & Geomech. Abstr.* **29**(3), 279-292.

Zheng, Z., Kemeny, J. and Cook, N.G.W. 1989. Analysis of borehole breakouts. *J. Geophys. Res.* **94**(B6), 7171-7182.

Zoback, M. L. 1992. First- and second-order patterns of stress in the lithosphere: the World Stress Map Project. *J. Geophys. Res.* **97**(B8), 11761-11782.

Software information

Ordering information – In order to allow users of this book easy access to the programs described in the text, basic versions of the programs ***Dips***, ***Phases***, and ***Unwedge*** are being made available through the Rock Engineering Group of the University of Toronto. These versions have all the functionality needed to follow the examples in the book.

If you are interested in receiving information on the advanced versions of these programs or on the other rock engineering programs which are available for distribution, please check the box on the order form and an information sheet will be mailed or faxed to you.

Placing your order – If you are ordering from outside Canada, fill in the 'International Order' section. The shipping/handling charge is fixed at US $15.00, no matter how many programs are ordered. If you are ordering from within Canada, fill in the 'Canadian Orders' section. The CAN $10.00 shipping/handling charge is also fixed.

Methods of payment – For international orders, a cheque or money order in US dollars, drawn on a US bank, can be used. You may also make your payment by credit card. The amount billed to your credit card account will be at the Canadian dollar equivalent, based on current exchange rates. For Canadian orders, a cheque or money order, drawn on a Canadian bank may be used. Also, the credit card payment option is available.

Shipping information – Please type or print your name, address, phone and fax numbers in the space provided on the order form. Mail or fax a copy of the order form to:

Rock Engineering Group, 12 Selwood Avenue, Toronto, Ontario, Canada M4E 1B2, Fax number: 1 416 698 0908, Phone number: 1 416 698 8217.

ORDER FORM

International orders

	US $
☐ ***Dips*** - $ 50	
☐ ***Phases*** - $ 50	
☐ ***Unwedge*** - $ 50	
Shipping/handling	$ 15.00
Total	

Payment by cheque or money order

Please make cheque payable to :

Geomechanics Program Account

Payment by Credit Card

☐ VISA ☐ MasterCard ☐ AMEX

Account number: ____________________

Expiration date: ____________________

Name on card: ____________________

Signature: ____________________

Canadian orders*

	CAN $
☐ ***Dips*** - $ 65	
☐ ***Phases*** - $ 65	
☐ ***Unwedge*** - $65	
Shipping/handling	$ 10.00
Subtotal	
7% GST (of subtotal)	
8% PST (of subtotal, Ontario only)	
Total	

* Orders must be shipped to a Canadian address

Shipping information

Name:

Address:

Country: Zip Code:

Phone: Fax:

☐ Please send information on advanced versions of programs and other rock engineering software.

Author index

Subject index

Printed in the United States
by Baker & Taylor Publisher Services